Archibald Smith

Observations on the Inca and Yunga Nations, Their Early

Remains

And on ancient Peruvian skulls

Archibald Smith

Observations on the Inca and Yunga Nations, Their Early Remains
And on ancient Peruvian skulls

ISBN/EAN: 9783337309671

Printed in Europe, USA, Canada, Australia, Japan

Cover: Foto ©berggeist007 / pixelio.de

More available books at **www.hansebooks.com**

OBSERVATIONS

ON THE

INCA AND YUNGA NATIONS,

THEIR EARLY REMAINS;

AND ON

ANCIENT PERUVIAN SKULLS.

BY

ARCHIBALD SMITH, M.D.

(LATE OF LIMA.)

From the Proceedings of the Society of Antiquaries of Scotland, Vol. V.

EDINBURGH:
PRINTED BY NEILL AND COMPANY.
MDCCCLXIV.

OBSERVATIONS

INCA AND YUNGA NATIONS.

WHEN, after three centuries of Spanish oppression, the Liberator General Simon Bolivar, restored the empire of Peru to the Peruvians, one of his officers observed regarding the ancient capital of the nation— " This city may with truth be called the Rome of America: the immense fortress on the north is the capitol; the Temple of the Sun is its Coliseum; Manco Capac was its Romulus; Viracocha its Augustus; Huascar its Pompey; and Atahualpa its Cæsar" (*Miller's Memoirs*, vol. ii. p. 195).

In the grand Temple of Cusco, before it was despoiled by the rapacious Pizarro, Almagro, and other merciless Goths, there were seen, seated on golden chairs, the embalmed bodies of the deceased Inca emperors, upon each side of a radiant figure of the Sun in massive gold. In the same sanctuary there was a chapel allotted to the image of the Moon—the acknowledged wife and sister of the Sun,—wherein the walls were lined all round with silver ornaments, as the compartment occupied by the image of the Sun was decorated with gold. On each side of the " Mamaquilla" or Mother-Moon, as here represented, the Stars were placed as her attendants; and the deceased wives of the Incas sat near her in the order of their respective seniority. Father Acosta compares this adoratory of Cusco—the ancient capital of the Incas—to the Pantheon in Rome, and remarks that the devil had persuaded those Indian infidels to erect to his service temples that in magnificence rivalled those dedicated to the true God.

Allured by the mild climate, beauty and fertility of the valley of Urubamba, in the vicinity of Cusco, the Incas there constructed baths and a superb country-seat, where they resided for the greater part of the year. They spared no cost or labour in embellishing with the treasures of Art at their command this charming abode of the imperial family. Not only did they build at this elysium sumptuous temples, stately palaces, artificial fountains, and extensive water-works with pipes or aqueducts of gold and silver; but, more than this, they made a decorative garden, with a zoological collection of animals from all parts of their wide-spread empire, which extended from Quito to the river Maule in Chile; and among the fanciful ornaments of their pleasure-grounds were flowers, shrubs, and fruit-bearing trees, all modelled or worked in the same precious metals, as we still see imitated on a small scale by the tasteful ladies and nuns of Lima.

If we descend from the Sierra to the coast of Peru, peopled by the Yunga nation, we shall find that, though the materials for building be different, the same solid grandeur of structure will strike our attention on viewing the ancient ruins of Pachacamac, situated on the elevated verge of the valley of the same name, commanding a wide prospect of the calm, unruffled ocean, and gorgeous sunsets.

These ruins, as they lately stood, consisted of the remains of a fortress, a palace, and a temple—all constructed of earth and sun-dried bricks. The palace is full half a league in circuit. The fort is on an eminence some hundred yards from the palace, and is a quarter of a league in circumference. It is constructed of three walls, broader than those of the palace; they are built in a terraced form, so that each receding wall commands and overlooks the one before it.[1] The temple has been a grand edifice, but unfortunately undermined and pulled down by hunters after treasure, or others led by idle curiosity to explore the graves, which are open in many places, and skulls with the hair perfectly adherent and intact are scattered about in profusion. The interior walls, covered with mud plaster, still exhibit rude paintings in red and yellow colours, such as we may call fresco paintings. Among these may be seen a kind of scroll, sometimes compared with the Grecian or Egyptian; and on the lower terrace, facing the sea, are the remains of decorated pilasters. History records that much of the heaps of treasure faithlessly exacted by Pizarro for the feigned ransom of the unfortunate monarch Atahualpa was from this temple of Pachacamac—(*Acosta*, vol. ii. ch. xii.)

In the reign of the ninth Inca, Pachacutec, his son and successor

[1] Ulloa—Noticias Americanas: Entretenemiento xx.

Capac Yupanqui, who ascended the throne in 1423, was General-in-chief of the army. This royal personage descended from the Sierra by the province of Yauyu to the valley of the Rimac, and is said by Garcilaso to have been the first of the Incas who saw the South Pacific. When he arrived on the coast he found the Temple of Pachacamac standing. It was built by the predecessors of Caysmancu, the Yunga king of the valley of the Rimac, &c.[1] Capac Yupanqui visited it, and we are told (*Garcil.*, vol. i. book vi. ch. 30) he entered this temple with silence and deep feelings of reverence, without the distraction or noise of prayers and sacrifices. The internal worship inculcated by the Incas was that of one great invisible Being, the creator and upholder of all things, whom they nominated Pachacamac (a word which means the Soul of the Universe); and the Sun was to be adored as his visible representative. But the Incas, to whom homage was rendered as the Children of the Sun, had never raised an adoratory, or devised an image or idol for the outward worship of their supreme God. It was even prohibited among them to utter the sacred name of Pachacamac, except under very special circumstances, and when spoken it was always with profound self-humiliation and contrition of heart. The politic Capac Yupanqui, as he approached the headlands of the valleys of Rimac and Pachacamac, sent envoys of peace before him, urging upon the Yunga king, Caysmancu, the unsuitableness of two peoples who acknowledged the same supreme God of the universe going to war with each other, and requesting the suspension of hostile measures on either side until they should have the opportunity of personally talking over their differences. Caysmancu admitted this reasonable proposal. The interview of the contending parties ended amicably. Casymancu agreed to acknowledge the supremacy of the Incas, under whom he was to continue to govern his own dominions, and to introduce among his people the worship of the Sun as the visible representative of the Great Spirit who ruled the universe.

The Yunga nations did not think it derogatory to their public worship of Pachacamac to introduce into the temple dedicated to his service

[1] The word Yunga or Yunca is of Quichua origin, and means "the hot valleys or territories," and was therefore applied to the inhabitants of the Peruvian coast. The Indians occupying the warmer regions of Eastern Peru and Bolivia are sometimes called Yungas, or people of "tierras calientes," without reference to any similarity of race.

images or idols, which were principally figures in imitation of fish of different sorts, significant of the bountiful supplies of food from the ocean ; a subordinate form of idol veneration or adoration which was common to all the Yunga race of the Peruvian sea-board—(*Garcilaso*, vol. i. book vi. ch. xvii.) The Inca Yupanqui removed those finny idols from the Temple of Pachacamac, which he enriched and adorned, and forbad the Yungas to offer up human sacrifices, as they were accustomed to, at their great festivals. Garcilaso repudiates the Spanish notion—originating from a misapprehension of the Quechua language— that the Incas offered up human victims in their temples, asserting that the only human sacrifices ever made under their benign institutions were at the interment of their sovereigns, or of some great curaca of their nation. He admits that upon those occasions the attached wives and faithful servants of the deceased eagerly and voluntarily pressed forward to be swallowed up in the tomb of the departed, actuated by the firm belief that after this life they were to survive in another state of corporeal existence; and this was the reason why they bestowed so much care on preserving and embalming their dead—(*Garcilaso*, vol. i. book vi. ch. 5.)

The dominion of Caysmancu extended northward as far as the river Pativilca; and beyond that border line were the possessions of the Grand Chimo, extending in the direction of Trujillo.

The Inca Yupanqui crossed this *Rubicon* with a formidable army, demanding allegiance to the sovereign rule of the Inca and the worship of the Sun, "who shone alike on all his creatures, and daily made the revolution of the world to behold their wants and supply their necessities." The Grand Chimo nation wanted no innovations, and they defended themselves gallantly against the unprovoked invaders of their independent territories and religious usages. But after a fruitless struggle, the Grand Chimo surrendered to the clemency rather than to the arms of the Inca, who treated him with great distinction, loaded him with gifts—especially supplies of clothing,—and left him to reign as the curaca or governor of his own people and territories, subordinate to the Inca as his sovereign.

The palace of the Grand Chimo—a massive building of small sun-dried bricks and mud—is still to be seen on the shores of the Pacific, between 7 and 8 degrees of south latitude. The whole district around this ancient monument, from Trujillo to Lambayeque, is crowded with the remains of towns and villages partly buried in sand, and sadly damaged by the searchers for gold and silver utensils of bygone generations, while the less coveted fictile remains of the potter are passed over with comparative indifference.

To Mr Waddell Blackwood of Trujillo I am indebted for the samples of tubulated pottery now before us;[1] and he informs me that the ruins of the Grand Chimo Palace are 400 yards in circumference. It has open areas within its walls, which, in such parts as are left standing, are yet about 30 feet high. Among the ruins are several water reservoirs, one of which is 40 feet square and 60 feet deep, all faced with stone; another from 40 to 50 feet broad, by 200 feet long.

After the conquest of the coast regions, of which the soil and climate, as well as the habits, customs, and character of the people, were materially different from those of the Sierra, the Incas introduced many great changes. They constructed spacious roads and astonishing aqueducts, by means of which traffic was facilitated, and agriculture vastly extended over the desert and sandy valleys of the coast. They raised up huacas or adoratories all over the plains; some of which even now, in a spoliated and disfigured state, stand out to the wondering gaze of the modern traveller as so many monumental *Calton Hills*, overlooking the shattered remains of Inca and Yunga civilisation, which everywhere strike the eye along the maritime valleys of Peru.

In the district of Chucuito, upon the borders of the great lake of Titicaca, at the elevation of 12,725 feet, there are—says Garcilaso—some vast edifices, among which there is a court or "patio," of 30 yards square, with an inclosure twice a man's height. On one side of it is a saloon 45 feet long and 22 feet broad. This *patio*, with its walls, floor, saloon, gateways, lintels, and roof, are all excavated in one solid rock. The outer roof, which is also an integral part of the same rock, is made in imitation of a covering of straw, or thatch, such as the Indians cover their houses with in the Sierra. The native tradition is, that this great rock palace or temple was dedicated to the "*Maker of the universe.*" Close to this excavated monument of immemorial antiquity, there are a great many sculptured stone figures of men and women, reported to be so lifelike as to appear actually living. They are represented in different atti-

---

[1] The basket of wicker with the open-mouthed painted drinking cups are from Pisco, the celebrated guano district of Peru (since presented, with various other relics, human crania, &c., to the Museum of the Society).

tudes and positions—some sitting, others standing with vases in hand, as in the act of drinking or stepping over a streamlet which runs through the premises. Other statues, of the same kind, represent mothers with infants in arms, or in various other attitudes. The local tradition is, that on account of some heinous sins committed, and especially for having stoned a man who passed through that district, the offenders were themselves converted into those statues of stone.

From Garcilaso we further learn, that Mayta Capac, the fourth Inca, stimulated by the ambition of his predecessors, to extend the boundaries of his empire, together with the special idolatry of Sun-worship, invaded with a chosen army the province of Tiahuanacu, on the south-eastern borders of the lake Titicaca; thus marking the comparatively modern era of the Inca occupation of that celebrated seat of pre-historic architectural remains.[1]

Among its ancient monuments, there is a large made mountain, reared upon a foundation of massive stone-work. Apart from this stupendous monument there are two gigantic figures of men, carved in hard rock or stone, with long vestments that reach the ground, and ornamental head-dresses, all bearing the appearance of the tear and wear of great antiquity. The city of Tiahuanacu was surrounded by a wall built of immense stones; and it is inconceivable how they could have been placed in *position*, or conveyed to that spot, for there are neither rocks nor quarries near it. Among the ruins of this old city are also to be seen many sumptuous frontispieces of houses, perfectly entire, and hewn out in one solid block of rock. Many of these façades look as if they only stood on detached masses of stone, some of which, by measurement, are 30 feet long, 15 broad, and 6 feet in front; and yet, when closely examined, they are all found to be of one piece.

The native Indian tradition is, that all those edifices, and others not enumerated, were works constructed anterior to the epoch of the Incas; and that it was in imitation of them that the fortress at Cusco was built in later times. (*Garcilaso*, vol. i. book iii., and chap. 1.)[2]

The largest island in the Archipelago of Titicaca has long been held sacred by the Peruvians, from the popular belief that it was on it that,

---

[1] Mayta Capac ascended the throne in 1171, and died at the age of ninety-two, in the year 1211.—Urrutia " Epocas del Peru."

[2] This great fortress of Cusco was built by the Inca Yupanqui, who reigned thirty years, and died in his seventy-ninth year of age, in 1453.

about the middle of the eleventh century, Manco Capac and his wife, Mama Oella-huaca, descended from the sun. Skirting the shores of this great water-basin of the Andes, we still see, as designated on our maps, not far from the peninsula of Chucuitu, the Isthmus " Yungillo," which is merely the diminutive of Yunga, and may, perhaps, have been so named as commemorative of this spot being the early cradle of the Yunga family. It certainly appears to be a fact indicative of a much earlier civilisation than can be ascribed to the period of the Inca occupation of those places on the shores of Titicaca, that the rock-sculptured temple at Chucuitu was dedicated to the " Maker or Creator of the Universe"—to whom the Incas, as we have already noticed, neither raised temples nor offered public worship. On the other hand, the Yungas of the coast, who had not, like the Incas of the Sierra, set up the sun as the emblem of either civil or religious dominion, bowed their heads before the Supreme Creator and upholder of all things, in their own great temple of Pacha-camac; thus preserving, in their outward worship (until subdued by the Incas), more truly than their contemporaries beyond the western Cordillera barrier, the memory of a higher religious life. But as the Inca and Yunga nations essentially agreed in acknowledging the same sovereign god and creator, Pachacamac, as did likewise (by Garcilaso's report of native tradition) the primitive founders of the more ancient rock-temple in Chucuitu, we may be, not unreasonably, permitted to infer that the progenitors of both the Incas and the Yungas made their first exodus from among the people who, from time immemorial, inhabited the borders of the great inland lake of Titicaca.

Now, admitting the affinity of origin and race of the collective indigenous tribes of the coast, and the Sierra of Peru, before the arrival of the Spanish invaders, we may expect to find among them all a certain family likeness. And this is the case. As a whole, they are of short stature, with small compact hands and feet; but on the mountains, where the air is highly rarified, we find a proportional expansion of lung and depth of chest, a ruddy complexion, and remarkably firm, well-turned, muscular limbs. In colour of skin, they vary considerably, according to elevation and climate, individual constitution, and other causes I have seen in the valley of Huanuco an industrious agricultural Indian family, of the name of Avila, who were distinguished by a fair complexion and lightish hair; and I have been told of tribes on the eastern frontiers

who are said to be nearly as white as Europeans.   But making allowance
for every subordinate divergence of colour, the prevailing tint of the skin
is brownish, though in some instances it deepens into bronze, and in the
humid sultry forest land verges to yellow.

The hair of the head is usually black, and even in old age rarely turns
to grey.   When allowed to grow long, as in the female, it is seen to be
straight, coarse, thickly set, and superabundant.   Some of the men have
thin beards, but for the most part they are smooth-faced like women.
In the physiognomy, size and shape of the face, we meet with consider-
able variety.   From the interior of South Peru, I have seen men with
strongly marked Mongolian features, and also not a few with the Jewish
profile.   In some natives of North Peru, again, I have been accustomed
to observe a broader and less aquiline nose than we so frequently notice
in the ranks of Indian infantry from the warrior departments of Puno
and Cusco.   In fact, there is much ground for believing, that in these
southern regions of the Andes, an intrusive race mingled with the primitive population at some unknown stage of their history.   One of the
provinces of the department of Cusco is named Aymaraes, and in the district of Titicaca—Puno, Huancane, &c.—the Aymará is currently spoken.
This language is allowed to be distinct from the Quechua, as well as its
cognate dialects, the "*Lumana*" of North Peru, the Chinchaysuyo, and
the Yunga.[1]   The coast lands of Nasca and Chincha, &c., which were
peopled by the Yungas, were included in that section of the Inca Empire
which, to the south, extended inland from Chincha to Ayacucho, and was
called Chinchaysuyo.   Among the books sent to Lima in my time, under

---

[1] When the tenth Inca, Yupanqui, took possession of the Yunga district of Nasca,
on the Pacific, he from thence colonised the corresponding hot valley of the Apurimac, and thence again it is most likely the banks of the Amarumayu, in the province of Mojos, inhabited by the Chuncho Indians.—Garcil., vol. i. book iii. chap.
xix. ; and also book vii. chap. xiv.   When one of Yupanqui's enterprising predecessors, Roca the sixth Inca, who died in 1363, occupied the throne, he sent his son
and successor Yahuar-Huaca across the Antis or Eastern Cordillera, with a large
military force, to the conquest of the country, afterwards called Antis-suyo.   On
the eastern side of the Antis he founded the colony of Pillcupata " de gente advene-
diza ;" that is, he peopled it with strangers from some other warm region of the
empire, for it was the practice of the Inca Emperors never to colonise a hot district
with natives of a cold alpine climate, and *vice versa*.   See Garcilaso, vol. i. book iv.
chap. xvii.

the auspices of the London Tract and Missionary Society, was the Gospel according to St Luke, translated into the Aymará, by Don Vicente Pasos-Kanki, a native of Cusco. I showed a copy of this translation, with the Spanish annexed, to a competent judge, who read and spoke the language; he read some verses in my hearing, but concluded by saying, it was impossible to express correctly all the labial, dental, and guttural sounds of the Aymará by any combination of the letters of the Latin alphabet. Just as the Gaelic is used by our Highland clergy, the curates of the Andine regions of Peru conduct the services of public worship and the instruction of the confessionary through the medium of the Quechua. This ancient language does not offer the same difficulties of pronunciation as the Aymará; but it has only in legitimate use eighteen letters of the Latin alphabet, it excludes the letters B, D, F, G, I, X (so that, according to its native structure, the word "Inca" should be written Ynca, and the word Yunga should be written Yunca); but this rule of Indian orthography has not been duly attended to by Spanish writers. I have never seen a Yunga Vocabulary, though I believe a Grammar, with some specimens of that dialect, in the "Confesionario," has been published by Fernando de Carrera. I learn, however, from the Vocabulary of the Chinchaysuyo, by Juan de Figuredo, that the principal difference between this Peruvian dialect and the Quechua consists in a grammatical *syncopa*, or cutting off a letter or syllable from the Quechua; as for example, instead of micurcani, micurcà; instead of munarcani, munarcà, in both which instances the particle *ni* is cut off in the Chinchaysuyo.

By the testimony of language we thus trace the affinity of race among the great bulk of the Peruvian nation under the Inca dynasty, while the isolated Aymará, still preserved and spoken in the very centre of that great empire, remains as a subject of curious inquiry. On the coast too, near Lambayeque, in the ancient dominions of the Grand Chimo, we have, in the town of Eten, a peculiar people who speak a language unknown to the rest of Peru; it is said to be a dialect of the Chinese.

But if we look into the ancient Huacas, or the more humble domestic vault of the poor Inca Indian, we further meet with proof of a general similarity of race in the cranial conformation, which may be considered as characteristic. Among the prominent signs may be enumerated,—a naturally low and narrow forehead, as compared with the interparietal, or lateral swell; a short longitudinal diameter; and, very commonly, a more or less depressed occiput. But that the Indian forehead is not always naturally low, a striking evidence is found in the portrait of the late Archdeacon of Cusco, the much-honoured Dr Justo Sahuaroura, who was the last of the Incas of Peru.[1] Intellectually, the educated Indian of Peru is allowed to be quite equal to the white Creole.

I do not propose to speculate on the causes or the consequences of the reigning characteristics of the Peruvian crania, nor enter upon details regarding the samples of ancient art manufacture found in private graves or Huacas, but shall cursorily notice a few particulars in explanation of the crania now exhibited. The skulls marked 1, 2, 3, 4, are from two of the most ancient Indian ruins in the valley of the Rimac; namely, the Huaca of Salinas, and the old city of Cajamarquilla, a few leagues to the north-east of Lima. Here (or at the neighbouring Huaca of Late), it is supposed that the speaking oracle "*Rimac*," after whom this valley is named, received deputies, and through their instrumentality, presided over the fate of neighbouring nations (Garcilaso, book vi. chap. vi. 24, 30, vol. i.)

If we bear in mind that this valley was not under the jurisdiction of the Inca government, until the era of Capac Yupanqui, who succeeded to the monarchy in 1423, and that Pizarro founded the city of Lima in 1535, we have pretty sure data for concluding that four of the Peruvian skulls now before us, viz. Nos. 1, 2, 5, 6, belonged to living men of the period intervening between these given dates. They were taken from their respective burying places, distinguished with the true sepulchral insignia of the Inca dynasty; the wrappings, and woollen textile materials, yarn, and other articles of ordinary use, of which I present samples from their own tombs. The Yungas required but little clothing in the warm climates of the valleys of the coast,—they had not made much progress in spinning or weaving before the Inca invasion of the coast, and their clothing was of cotton or soft pliant grass; the Vicuña wool is a sure indication of the Inca manufacture and rule. Had they died before the epoch of the Inca occupation of the coast, they would not have these sepulchral accompaniments—had they died while under Spanish rule, they would have had Christian burial in consecrated ground; whereas the specimens before us were carefully procured from the most undoubted Indian graves. Nos. 3

[1] Sahuaroura witnessed the final triumph of his countrymen at Ayacucho in 1824. He wrote his " Recuerdos de la Monarchia Peruana" in 1836, which was published, with portraits of the Incas in Paris. An. Dom. 1850.

and 4 may be anterior to the Inca period, as they were exhumed without clothing. In the valley of the Rimac, the dead are found in neatly constructed and plastered underground cellars, beneath the ground floor of ancient native dwellings, and the Huacas, that appear like so many made mounds, widely dispersed over the plains, have their mausoleums which can only be entered by very low apertures.[1]

As to the forms of the skulls on view, it may be seen at a glance that they vary considerably. Nos. 1, 2, 3 and 4, were picked up as they came to hand, without selection; but Nos. 5 and 6 were chosen to illustrate two varieties of conformation, which, though not so common in the graves, are often alluded to in books. Thus, No. 5 is notable for its very slanting forehead, and compressed occiput at the parietal protuberances; while No. 6 is an exceedingly well formed skull. They were both taken from the old burying-ground south of Chorrillos, on the way to the ruins of Pachacamac from Lima. This No. 6 exhibits the peculiarity of a triangular or wormian super-occipital bone, to which, in the " Edinburgh New Phil. Journal," January 1858, Dr Daniel Wilson alludes, as being, according to Dr Tschudi, " *peculiar to the Peruvians, and traceable in all the skulls of that race.*" If Dr Tschudi be here accurately quoted by Dr Wilson, there cannot be a greater mistake on a question of fact always open to be decided by personal observation in Peru. The result of my own recent investigations in reference to it, when in Lima, during the years 1859 and 1860, I have stated elsewhere;[2] and shall only now remark, that, as far as the Peruvians are concerned, the existence of wormian bones are not usual cranial signs, and cannot be relied upon as of typical significance.

With regard to the oblong, flat crania, which have been occasionally found around the lake of Chucuito, or Titicaca, on the Andes, a region

[1] The ancient graves in the province of Tacna present a remarkable uniformity. All are arranged in parallel lines, and in shape of small subterraneous arches, under which the bodies are interred, all of them in the same position, accompanied with the instruments or implements, &c., used by them when in life. And in the valley of Palca, as well as in that of Caplina, and some others, are yet to be seen enormous granitic stones covered with engraved hieroglyphics and curious figures,—works, perhaps, of another race, and of more advanced civilisation, who knew the use of signs in preserving the chronicles of events.—*La Revista de Lima*, 15th March 1863.

[2] See " Edin. New Phil. Journal, New Series," vol. xi. 1860; Art. " Peruvian Gleanings." By Dr Archibald Smith.

to which there is easy access from Tacna and the coast district of Arica, where such skulls have been exhumed in greater numbers, I cannot speak from personal experience. It has indeed been suggested by Dr Unanue of Lima, that there is an analogy between the Malay and Aymará languages, and that intruders of the Malay race may have landed from their boats or balsas at the harbour of Arica, and penetrated to the Sierra by the usual Cordillera pass. But this is only one of many conjectures on the subject of intruders, which throws no additional light on the origin of the flat skulls. In 1859, I saw in the Lima Museum two remarkably deformed crania, one of which pertained to an encased mummy, so strongly compressed on the forehead, that in life the bulk of the brain must have been pressed back on the occipital region ; but these were singularly exceptional specimens.

There is no accounting for the false taste of barbarous nations, as exhibited in their conventional usages.

Garcilaso tells us of the practice of the savage " *Mantas* " of Esmeralda, whom he visited on his voyage from Peru to Spain, in the year 1560. The adults delighted in covering their faces with scars and daubs of yellow, blue, red, and black paint, all blended differently according to the taste of each individual. These people pressed their children's heads between two boards, one on the forehead and the other on the occiput, which, day by day, they drew tighter and tighter together, until the children attained the age of four or five years. By this treatment, our historian says, the head was made broad laterally, and narrow longitudinally.

The late Lieutenant Herndon of the United States, in his work, entitled " Exploration of the Valley of the Amazon," published in 1853, relates, at page 203, an example of the same barbarous custom, which he witnessed among the " Conibos" of the Ucayali,—" The head of the infant had been bound in boards, front and rear; and was flattened and increased in height."

NOTE.—In reference to the barbarous custom of the Conibos "*de aplastar la cabeza de sus niños con dos tablas*," that is, of flattening their children's heads between two boards, A. Raimondi, Professor of Natural History in the Medical School of Lima, relates a curious instance observed by him during his recent travels in the regions of the Ucayali and Marañon, in the province of Loreto, as follows :—

" In the missionary station of Sarayaco I had the opportunity to see a male child—"un niño"—which its mother had brought in order to be baptised, whose head was elongated backwards, while there was a rounded protuberance on the well depressed frontal bone. Not comprehending how this protuberance could have grown up on a part compressed by a board, I asked the mother if the board she had used to compress this child's head was a flat one, and learned that the board had a large hole in it, which explained how the protuberance on the frontal bone corresponded with the opening in the board, and developed itself where that bone was free from compression." Professor Raimondi remarks, that by the artificial compression of the Conibos children's heads, the forehead is forced to recede, while the skull is lengthened backwards, and thus they much resemble some skulls found in certain "*Huacas*" or ancient burying places in the Sierra. (See "Revista de Lima" (1862), article "Apuntes sobre La Provincia Litoral de Loreto," by Antonio Raimondi.

I may just observe, in reference to the ancient occupants of Peru, that I think very little is yet generally known of the antiquarian remains of their civilisation before the Spanish conquest. I have had free access in Lima, some years ago, to a magnificent private collection of about three hundred drawings from ancient monuments and ruins of cities—not less interesting in character, I believe, than those of Yucatan—which were beautifully executed by Mr J. Raymond Clarke, an American gentleman who had devoted many years to such researches among the least frequented forest lands of Peru, and other neighbouring republics.

The following Donations were presented to the Museum by Dr Smith.

Collection of Remains from the Ancient Tombs of the Inca and Yunga Nations in Peru, including :—
Human Skull from the ruins of the Huaca De Salinas, near Lima, Peru.
Three Human Skulls from the ruins of Cajamarquilla, near Lima.
Two Human Skulls from the ancient Huaca, Chorrillos, Lima. One of these crania shows the existence of a wormian bone, it is of large size, and is situated at the junction of the sagittal and occipital sutures. Several smaller distinct ossiculæ are also present in the occipital suture. Dr Smith states that wormian bones are by no means general or characteristic of the cranium of the ancient Peruvians.
Various portions of the Dresses of Cotton and of Woollen Cloths, some plain, striped, checked, and others with patterns in different colours;

also Ornamented Belts, Bunches of Thread, &c., &c., found in the Ancient Tombs of Peru.

Basket formed of reeds and plaited grass, containing two cups, being portions of gourds, balls of very fine woollen thread, and wound on three wooden spindles; Short Head of Maize or Indian Corn; Ring-shaped Stone, 3½ inches in diameter, with a perforation in the centre 1¼ inch in diameter; and Small Stone Ball 1½ inch in diameter, found in a grave at Pisco.

Two Small Idols, male and female, of black earthenware, from Trujillo.

Four Bottles of Red and Black Earthenware; one representing a monkey, another twin birds, &c., from Trujillo.

Two Drinking Cups, of thin, dark-coloured earthenware, from Pisco, and shaped like the modern tumbler.

Two long shaped Drinking Cups, rounded below with bell-shaped mouths, of reddish earthenware, with coloured patterns; one measures 7 inches in height; on it is painted the cactus or melon thistle, in black and white; the other is covered with a rich pattern of red, white, and black colours, from Pisco.

Bronze Star-like Implement, of five rays, one of which is awanting, from Pisco.

Small-sized Face or Mask, in red earthenware, of an Indian, displaying in black colour a thin beard and moustaches, from an old Peruvian grave at Chincha.   The Peruvian Indian is usually beardless.   .

Portion apparently of the Spindle of a Distaff, terminating in a pear-shaped extremity, which is hollow, and is pierced at each end with a round hole, and a small piece of slab pottery, pierced in the centre, probably used with the spindle.

As specimens of the language of this ancient people :—

Padre Diego de Torres Rubio, Arte y Vocabulario de la Lengua Quichua general de los Indios de el Peru—to which el P. Juan de Figueredo added a Vocabulary of the Chinchaysuyo, printed in Lima, 1754, 12mo.

El Evangelio de Jesu Christo segun San Lucas, en Aymará y Español, 12mo, Lond., 1829.

PRINTED BY NEILL AND COMPANY, EDINBURGH.

ON THE

SPOTTED-HÆMORRHAGIC YELLOW FEVER

OF THE

PERUVIAN ANDES,

IN 1853-57.

By ARCHIBALD SMITH, M.D.

[REPRINTED FROM THE "TRANSACTIONS" OF THE EPIDEMIOLOGICAL SOCIETY.]
of London

ON THE SPOTTED-HÆMORRHAGIC YELLOW FEVER OF THE PERUVIAN ANDES, IN 1853-57.

By ARCHIBALD SMITH, M.D.

(*Read December 2nd*, 1861.)

Introduction.—On the 20th of March, 1860, I had the honour of being nominated by the Medical Society of Lima, as one of four physicians on its Epidemiological Committee. I was thus placed in a position, which led me to inquire with more care into the origin and progress of certain epidemics in Peru, and very particularly of the yellow fever, which reigned on the Andes since the latter end of 1853.

It is supposed by some medical men in that country, that the epidemics on the coast and in the mountains at this time were only coincident pestilences, without any essential similarity of nature or origin. Whether the peculiar fever which reigned on the coast in 1852-53, was of foreign or native growth, is a question which has been keenly discussed by the members of the Medical Society of Lima, but by no means satisfactorily settled.*

I believe the first case of decided yellow fever with black-vomit during the coast epidemic of 1853, occurred in a patient of my own, who died on the 16th of April. This was a lady from Arequipa. In the first consultation on this case, I stated to my colleagues, that it presented the symptoms of a true typhus-icterodes. All were agreed in looking upon it as an isolated case of this type ; but we could not ascertain how it had originated, though it was true that the lady resided at the time of her illness in a great commercial house, which was the very centre of communication with Atlantic ports, and resorted to by nearly all the ship-masters in the guano trade. It was not until a month after this melancholy event, several cases more or less of a similar character, but otherwise unconnected, were brought into public notice by a zealous and active prefect—General D. Pedro Cisneros. Thus, a new light began to dawn on the minds of those physicians under whose observation some such novel cases had fallen ;

---

* It may be deserving of notice that, from the year 1851 to 1859, Lima was subject to a succession of epidemics, ending in diphtheria and small-pox in 1858-59.

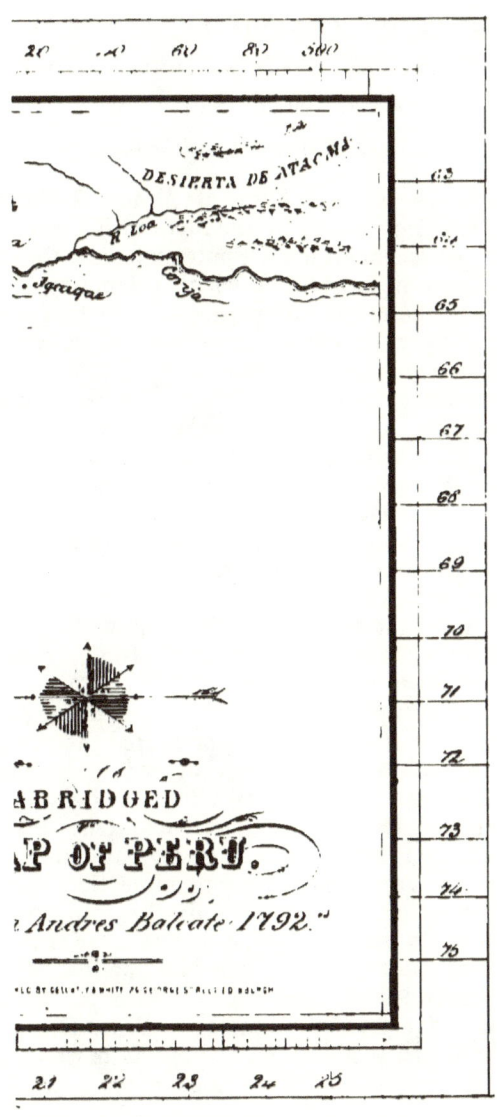

DESIERTA DE ATACMA

R Loa

Iquique

ABRIDGED
MAP OF PERU.
Andres Baleate 1792.

ON THE SPOTTED-HÆMORRHAGIC YELLOW FEVER OF THE PERUVIAN ANDES, IN 1853-57.

By ARCHIBALD SMITH, M.D.

(Read December 2nd, 1861.)

Introduction.—On the 20th of March, 1860, I had the honour of being nominated by the Medical Society of Lima, as one of four physicians on its Epidemiological Committee. I was thus placed in a position, which led me to inquire with more care into the origin and progress of certain epidemics in Peru, and very particularly of the yellow fever, which reigned on the Andes since the latter end of 1853.

It is supposed by some medical men in that country, that the epidemics on the coast and in the mountains at this time were only coincident pestilences, without any essential similarity of nature or origin. Whether the peculiar fever which reigned on the coast in 1852-53, was of foreign or native growth, is a question which has been keenly discussed by the members of the Medical Society of Lima, but by no means satisfactorily settled.*

I believe the first case of decided yellow fever with black-vomit during the coast epidemic of 1853, occurred in a patient of my own, who died on the 16th of April. This was a lady from Arequipa. In the first consultation on this case, I stated to my colleagues, that it presented the symptoms of a true typhus-icterodes. All were agreed in looking upon it as an isolated case of this type ; but we could not ascertain how it had originated, though it was true that the lady resided at the time of her illness in a great commercial house, which was the very centre of communication with Atlantic ports, and resorted to by nearly all the ship-masters in the guano trade. It was not until a month after this melancholy event, several cases more or less of a similar character, but otherwise unconnected, were brought into public notice by a zealous and active prefect—General D. Pedro Cisneros. Thus, a new light began to dawn on the minds of those physicians under whose observation some such novel cases had fallen ;

---

* It may be deserving of notice that, from the year 1851 to 1859, Lima was subject to a succession of epidemics, ending in diphtheria and small-pox in 1858-59.

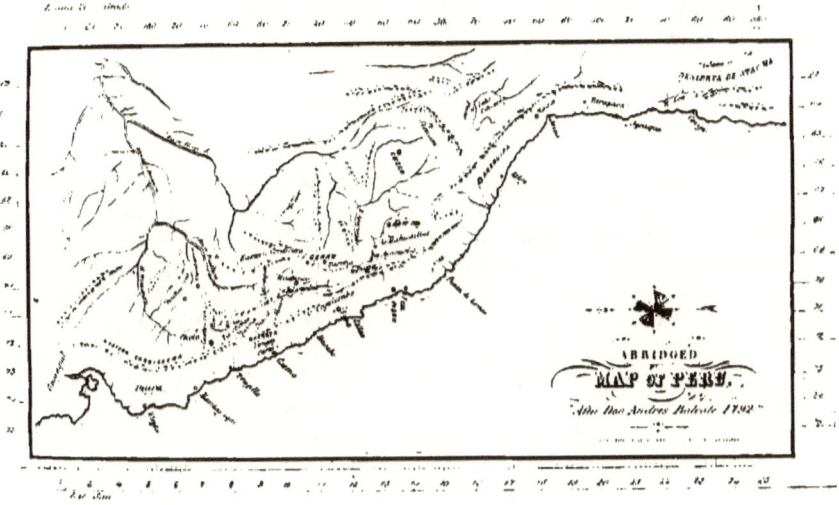

ABRIDGED
MAP OF PERU.
by
Allic Don Andres Baleato 1792

ON THE SPOTTED-HÆMORRHAGIC YELLOW FEVER OF THE PERUVIAN ANDES, IN 1853-57.

By ARCHIBALD SMITH, M.D.

(*Read December 2nd*, 1861.)

Introduction.—On the 20th of March, 1860, I had the honour of being nominated by the Medical Society of Lima, as one of four physicians on its Epidemiological Committee. I was thus placed in a position, which led me to inquire with more care into the origin and progress of certain epidemics in Peru, and very particularly of the yellow fever, which reigned on the Andes since the latter end of 1853.

It is supposed by some medical men in that country, that the epidemics on the coast and in the mountains at this time were only coincident pestilences, without any essential similarity of nature or origin. Whether the peculiar fever which reigned on the coast in 1852-53, was of foreign or native growth, is a question which has been keenly discussed by the members of the Medical Society of Lima, but by no means satisfactorily settled.*

I believe the first case of decided yellow fever with black-vomit during the coast epidemic of 1853, occurred in a patient of my own, who died on the 16th of April. This was a lady from Arequipa. In the first consultation on this case, I stated to my colleagues, that it presented the symptoms of a true typhus-icterodes. All were agreed in looking upon it as an isolated case of this type ; but we could not ascertain how it had originated, though it was true that the lady resided at the time of her illness in a great commercial house, which was the very centre of communication with Atlantic ports, and resorted to by nearly all the ship-masters in the guano trade. It was not until a month after this melancholy event, several cases more or less of a similar character, but otherwise unconnected, were brought into public notice by a zealous and active prefect—General D. Pedro Cisneros. Thus, a new light began to dawn on the minds of those physicians under whose observation some such novel cases had fallen ;

---

* It may be deserving of notice that, from the year 1851 to 1859, Lima was subject to a succession of epidemics, ending in diphtheria and small-pox in 1858-59.

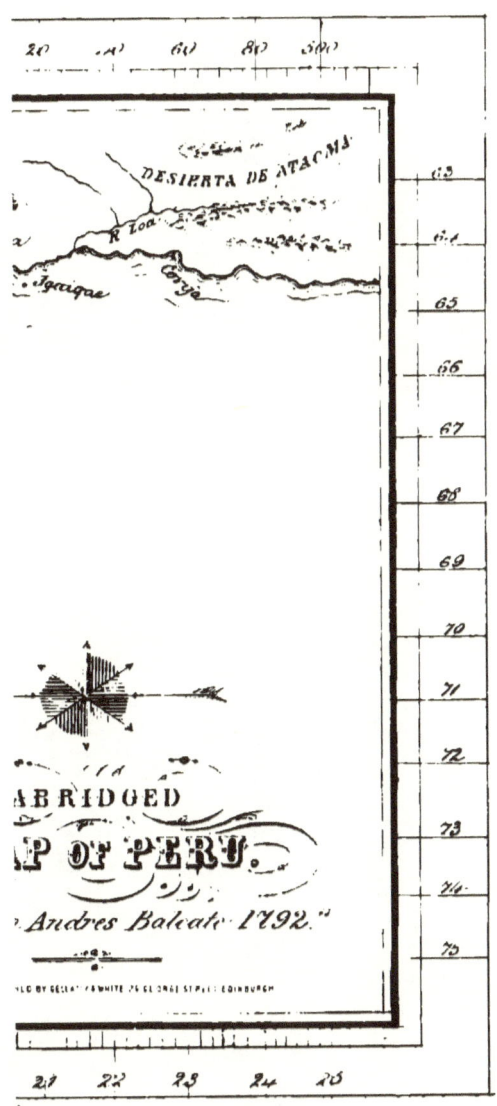

DESIERTA DE ATACMA

R Loa

Iquique

ABRIDGED

P OF PERU.

Andres Baleato 1792.

LC BY GELLA' & WHITE 26 GEORGE ST PLACE EDINBURGH

and then it was that, for the first time, the question was mooted—whether these exceptional cases were merely sporadic—and dependent on the autumnal change of season with its usual malaria, or whether they might not rather be looked upon as aggravated cases of the prevailing epidemic, then on the wane, but of which neither the yellow skin nor black vomit had been hitherto concomitant symptoms? According to the latter view of the question, the hitherto mild pyrexial disorders of 1852 and 1853, were only the primary and progressive degrees in the nascent and growing stages, of that mature yellow fever which burst forth with violence in January 1854.*

Another question much agitated in Lima has been—was the epidemic of 1852 and 1853, the offspring of infection or of contagion? By the former designation was understood the bad atmospheric condition of the locality, by the latter the mediate or immediate transmission of a morbid specific emanation from one body to another—from the sick to the healthy, in a locality not previously infected. In a limited sense, the contagious miasm of a diseased body allowed to accumulate in the sick-room, creates an infected local atmosphere; and, thus, in the English language contagion and infection are often used synonymously. On this question the arguments adduced by either party failed to convince their opponents, and the remark of La Martine (with respect to religious controversy) was applicable—" On ne prouve à l'homme que ce qu'il croit déjà." I can only say, that once the disease became general in the airy open houses of the capital of Peru, it was not easy to ascertain the precise mode of propagation, although it was certainly observed, that when the malady attacked one member of a family it generally spread throughout the household, affecting different individuals differently, according to race, or peculiar susceptibility.

Since the yellow fever pestilence of 1842-45, in Guayaquil, there had been repeated instances of persons who had had this fever on board the northern steamers, disembarking at Callao, and arriving in Lima.† But the fact, nevertheless, is a remarkable one, that it has only been since the discovery of gold in California, and the consequent rush of people and shipping

---

* See " Rise and Progress of Yellow Fever in Peru", Edinburgh Medical and Surgical Journal, vol. lxxxii.

† See the records of the Lima newspapers on the questions of local police, relative to yellow fever and its supposed importation from Panama, etc.—*El Comercio* of 17th April, 1852; *El Mensagero*, in its issues of 4th January, 17th March, and 18th May, 1853.

from the Atlantic to the Pacific shores, that these recent epidemics have sprung up in Peru.

Of the sanitary condition of numberless vessels that thus visited the Peruvian ports during the period referred to, it is now impossible to affirm anything with certainty. We know that early in 1852 the harbour of Rio Janeiro was crowded with shipping, and flags hoisted half-mast high as the signal of fatal yellow fever on board. How many of these vessels with their sickly crews were destined to the Pacific we have now no means of ascertaining, for every captain touching at Callao had an interest in not divulging any facts that might bring upon his ship the application of the quarantine laws. Yet, as the whole question is open to inquiry, I shall, by way of slight illustration, state one or two events in reference to it. About the middle of the year 1851, the American steamer, the *New World*, destined to California, anchored in the Bay of Callao, after having lost almost her whole crew of yellow fever in Rio Janeiro. On this subject, Lady Emeline Stuart Wortley informs her readers, that fifteen men belonging to the *New World* steam-ship died there, and the captain on his arrival in Callao, was ill from the effects of the severe attack he had had of that dreadful disorder.*

Another ship of the Pacific Steam Company, called the *Quito*, commanded by Captain Wells, entered the harbour of Callao, on the 8th of April 1852, having touched on her voyage from England at Rio Janeiro, where several of the crew caught yellow fever—a disease which had reigned there in a malignant form since early in January 1850.† The fever spread in the *Quito*, and nine men died of it before she had got to the Straits of Magellan. Among others, the commander of the steamer was attacked, and narrowly escaped by a fortunate accident. At his own suggestion he entered a bath, heated by steam. It poured into the bath so fast that it overpowered him. He fainted, and when rescued, was nearly parboiled. He was conveyed to bed. He soon recovered consciousness, and was bathed in the most profuse perspiration, which entirely carried off his fever.

One of the passengers aboard the *Quito* at this time, was a Limeñan, Don N. L., who was attacked with yellow fever, and confirmed black vomit. His case was to all appearance hopeless. But he happily survived, and when the ship made the port of Valparaiso, he was landed and left under medical

---

* Travels in the United States, etc., vol. iii, p. 200.
† Dr. J. O. McWilliam on the Yellow Fever Epidemy in Brazil, 1849-50.

treatment. On the 6th of May, this same gentleman arrived in Callao from Chile per steamer *Bolivia*, and on his arrival in Lima, he was often visited by one of his most intimate friends—the well known Mr. R—. This intercourse only lasted a few days, when the latter gentleman was seized with an illness, which, from the novelty and severity of its symptoms, he felt assured was the Rio Janeiro fever.*

Captain Wells, with whom I made my late homeward passage from Callao to Panama at the close of 1860, informed me in reference to the above events, that on his first arrival in the Bay of Callao with the *Quito* from England, in April 1852, he was allowed to land the passengers, and proceed, unmolested, as far as Panama. But on his return from the isthmus, he found the Peruvian functionaries at the seaport of Lima differently disposed. They expressed their intention of placing his ship—the *Quito*—in quarantine, on the express ground that in her recent passage from Europe she had been at Rio Janeiro, and came from thence an infected vessel. This purpose, however, was not carried into effect.

During the months of January and February 1854, when yellow fever raged in Lima, Madame Ida Pfeiffer visited that city and was the guest of M. Rodewald, the Hamburgh Consul. We may therefore, reasonably infer, that it was not without the most direct and reliable information received at the con-

---

* About this time Mr. R. was the importer of German emigrants into Lima. I was then led to believe that the cases of sickness and deaths of yellow fever, now ascertained to have been on board of the *Quito*, had occurred on board one of the emigrant ships. (Letter in *Lima Med. Gazette*, 15th July, 1859.) I thus lay under a misapprehension, which I am glad of this opportunity to dispel. We see that the source of this "Rio fever" was the *Quito*, and through this channel we have an instance of a tainted ship communication between the coasts of Brazil and Peru.

Humboldt writes (*Political Essay on New Spain*, vol. i) : "The *mattazahuatl*, a disease peculiar to the Indian race, seldom appears more than once in a century. It raged in a peculiar manner in 1545, 1576, and 1706." "We have no exact data as to the *mattazahuatl*. It bears certainly some analogy to the yellow fever or black vomiting; but it never attacks white people, whether Europeans or descendants from the natives. The individuals of the race of Caucasus do not appear subject to this mortal typhus; while, on the other hand, the yellow fever or black vomiting very seldom attacks the Mexican Indians. The principal site of the *vomito prieto* is the maritime region, of which the climate is excessively warm and humid; but the *mattazahuatl* carries terror and destruction into the very interior of the country, to the central table land, and the coldest and most arid regions of the kingdom." (P. 117.)

"A very interesting problem remains to be resolved. Was the pest which is said to have desolated from time to time the Atlantic regions of the United States, before the arrival of the Europeans, and which the celebrated Rush and his followers look upon as the principle of the yellow fever, identical with the *mattazahuatl* of the Mexican Indians?" (P. 118.) See *Note A.*

sulate, that this truthful lady states in vol. ii, pp. 135-6, of her *Second Journey Round the World*, that about two years previously not less than 2000 German emigrants were induced to leave their native land at the invitation of the government of Peru. "But the ships were overcrowded, the food and water bad, they were treated like the slaves brought from Africa, and more than the half of the unfortunate creatures died on the voyage."

What kind of disease prevailed so fatally in these ships we are not now able to ascertain, though it is not improbable that while in the yellow fever latitudes, a fever produced from pythogenetic causes generated on board might have assumed the typhus-icterodes type, which at that period reigned among the shipping along the coast of Brazil. Whether, indeed, any of the German emigrant ships in question had touched at all on the Brazilian coast, has, with conflicting, and therefore, so far, neutralised evidence, been controverted recently in Peru ; and official reticence casts an impenetrable veil over the fearful mortality on the passage so feelingly deplored by Madame Ida Pfeiffer. Certain it is, however, that the first of the German emigrant ships proceeding from Bremen arrived in Callao on the 9th of December, 1851, and early in January 1852, the mild adynamic or primary form of yellow fever broke out in Lima and the coast of Peru. This epidemic was vulgarly designated peste or "pelusa," just as in 1849 the same precursory form, under the name of "Polka," prevailed in Rio Janeiro—as described on the spot by Dr. Lallemant, and again quoted by Dr. McWilliam, as the first stage or stadium of yellow fever. I can state from my own observation that German emigrants were early sufferers in the Lima epidemic of 1852, which malady if not introduced by them, was at least contemporaneous with their landing in successive bands on the shores of Peru.

In July 1852, Mr. E. Gomes Sanches left Lima as envoy from the Peruvian Government to that of Brazil. On his arrival in Rio he related to some friends that a new kind of epidemic had just sprung up in Peru, and how he himself had been affected by it before his departure from Lima. Whereupon, he was told that such also had been the precursory form of yellow fever in Brazil, and that without doubt it would prove to be the same in Peru. A year after, Mr. E. Gomes Sanches returned to Lima, and soon found, as he informed me, this prediction verified to the letter. About the close of 1853, his young friend, Ensign (or Lieutenant) Don Melchor Caballero, was seized with yellow fever, and sent to the Military Hospital—"San Bartolome," where he had a narrow escape of his life.

Such facts as I have thus briefly noticed, indicate, I think, a striking similarity of preliminary nature between the growth of yellow fever in nearly corresponding latitudes on both sides of the South American Continent from 1849 to 1853-4.

THE SIERRA EPIDEMIC OF 1853-57.

It is estimated that the epidemic yellow fever, which for several years in succession, committed great ravages among the Andine population of Peru, proved fatal to full one-fourth of the indigenous race. The government endeavoured to arrest its progress by sending medicines and medical men to all the inland provinces, at an expense to the treasury of thirty-five thousand dollars per month.

An order was, moreover, issued to all the prefects of the departments to send by each post a statement of the numbers of sick, and the result of the medical treatment in the different lazarettos, &c. In the department of Cuzco the mortality was excessive, and in its capital alone out of a gross population of whites, Indians and mestizoes, variously estimated before the epidemic, at from 30,000 to 40,000 souls, 12,000 deaths are said to have taken place in the course of a few months.

The nature, and separate origin of the recent Sierra yellow fevers, I mean to consider chronologically, according to the dates of their outbreak, and the course they followed in the northern, southern, and central departments.

PART I.—THE EPIDEMIC OF CONCHUCOS 1853-55.

The first in point of time, is the epidemic, which commenced in September 1853, in a mild transandine climate, and in a village of agricultural labourers, called Yurma, with a population of about one thousand inhabitants. This is usually distinguished as the epidemic of Conchucos in the department of Ancachs. In September 1854, about one year after the commencement of the disease in Yurma, the Government of Peru, then agitated by civil war, had its attention drawn to the alarming spread of the malady in Ancachs. It was resolved by the authorities to send a medical commission to the capital of that department, Huaras. On their arrival at this destination, the commission were informed by the prefect of the enormous extent of the disease, and of the mortality that attended it.

They first endeavoured to ascertain the real character of the epidemic, by daily attendance, from the middle to the end of

September, at the Lazaretto of Huaras; and we learn from one of the commissioners, Dr. Daza, in a manuscript memoir on the epidemic of Conchucos, addressed to the prefect of the department of Ancachs, that the commissioners (four in number) agreed to classify the disease as a "typhus or typhoid fever."

After this recognisance each proceeded to his own allotted field of observation.

In Piscobamba this physician wrote a popular treatise on the epidemic, dated 26th February, 1855, which was sent to Huaras for publication and circulation. The following account of his practical observations in that temperate climate appears to me so carefully drawn up as to deserve perfect confidence. And as a guide in the history of this epidemic he further deserves priority, on the ground that it was in the province of Conchucos that the disease first became epidemic.

Signs of the First Stage in the Epidemic of Conchucos.— "The invasion is generally sudden, and commences with a general disorder of the system, accompanied with lassitude, disinclination for food, chills and heats; headache, sometimes confined to the forehead, or embracing the whole head; slight nausea, or vomiting of a green and yellow hue; pains in the thighs and loins; on rare occasions a lax state of the bowels. All this train of symptoms appears within the first twenty-four or thirty-six hours; but it sometimes happens that the failure in strength and appetite is experienced several days before the fever is established." [I may remark, that this prefatory feeling of indisposition for some days before a decided attack of yellow fever was likewise experienced, though not very often, I believe, in the Lima epidemic of 1853. I felt it in my own case in May of that year.]—A. S.

Symptoms of the disease when fairly established in its diverse forms—1. Of the Benignant, or simple Hæmorrhagic Form.—"The primary symptoms of invasion having passed, the patient borne down by the fever seeks repose, and presents the following signs of disease. Prostration of vital forces, greater or less in degree according to the intensity of the fever; the countenance is singularly changed; pulse at one time broad, full, and strong, and again hard and small, but always frequent—varying from 90 to 140 beats per minute; the heat of the body, dry and ardent, is much increased; the respiration frequent and from time to time the inspirations are deep and prolonged. So great is the restlessness in some cases, that the patients cannot remain quiet, but toss about and throw off the bed-clothes continually. The headache is at times very violent, and occasionally attended with a sense of heaviness and mental drowsiness. The eyes are loaded and watery like those of a drunken person ; the lips are somewhat dry ; the tongue is generally broad with a white spot in the middle, but red at its point and borders, at one time it is sharp-pointed and high coloured, at another broad, and covered with a white or yellowish tinge, and somewhat viscous. There is urgent thirst, no appetite, and the bowels are much confined. The urine is high coloured and almost red ; deep-seated aching pains are strongly felt in the loins, arms and thighs, being often, however, limited to only one of these parts.

"The symptoms enumerated go on increasing until the sixth, seventh, and eighth day, and are, occasionally, prolonged to the tenth or fifteenth day—accompanied, almost constantly, from the third to the fourth day with mulberry coloured spots, which present themselves on the skin : they vary from the size of a fleabite to that of a pea, and first appear on the lower part of the breast, afterwards on the abdomen—from whence they sometimes spread all over the body. About the same period, the patient feels in the epigastrium and right flank, a pain more or less acute when pressed upon, and also a gurgling noise, under the pressure of the finger, may be observed in the latter region.* In like manner, the belly becomes enlarged in certain cases, and after a time tympanitic ; but, in general, this is not the case, and the abdomen remains bland and yielding.

"During the earlier days of this fever, the patient is sleepless, or slumbers lightly, and now and then has mild delirium ; but he is, at other times, observed to fall into a profound sleep, from which it is difficult to awake him.

"Lastly, on the fourth or up to the seventh day, nasal hæmorrhage very generally breaks out in profusion, and not unfrequently demands the most serious attention.

"When the fever is to end in restoration to health it is observed that, from the eighth to the fifteenth day, it declines by degrees, as indicated by the moderate heat of the body accompanied with light perspirations, less frequency and greater softness of pulse ; free and easy respiration ; restlessness disappearing ; the headache declining into a mere sense of heaviness, with, at times, a peculiar noise in the ears ; the eyes

---

* Dr. Macedo, in treating of the epidemic fever of Ancachs as seen by h'm at Huaras, says : "This noise in the iliac fossa was a very rare symptom, and may be considered as an exception." *Lima Medical Gazette.*

becoming more expressive; the tongue gradually assuming its natural appearance; thirst declining; the appetite returning; the urine, usually more abundant. And thus, after ten or fifteen days convalescence, the patient recovers his strength.

"When the fever is to terminate fatally, then the pulse becomes progressively smaller, and more frequent; the breathing hurried; the heat of the surface from being moderate becomes ardent; the head is confused, and deafness often accompanies this symptom; the lips and tongue are dry—and, at times, the latter presents open chinks covered with an exudation of blood, or a dark stain; the voice fails—speech is broken and slow; the eyes are sunk, with a fixed look; the nose is sharp; the hands and feet are tremulous and bathed in a cold sweat; the pulse is no longer perceptible, and death closes the scene.

2. *The Bilious or Icteric Form.* —This typical form presents two distinct modifications—viz., the hæmorrhagic-icteric and the non-hæmorrhagic-icteric, or congestive. In the present instance, Dr. Daza describes the latter, which he designates 'the bilious.' "From the fourth to the sixth day— up to which time the course or symptoms are just the same as in the preceding (benignant variety), this bilious form begins to show itself by a greater prostration of strength, so that the patient is unable to move himself; the pulse is small and frequent; the heat of the body is dry and burning; breathing agitated; the voice weak, and the manner of utterance much altered; the skin of the face, neck, and breast, presents a colour which varies from an opaque yellow to the brightest tint of the broom-blossom. This yellow tinge is more especially conspicuous in the white of the eye, which, as the disease advances, grows more intensely yellow. The headache is ordinarily attended with a troublesome sense of heaviness; the tongue is covered with a white or yellow coat, and in continuation it becomes dry, arid, chapped, and covered with a dark or blackish stain, the same as the lips; there is a bitter taste in the mouth; retching and vomiting of a green and yellow colour; and such is the gastric irritability that it is generally necessary to apply a sinapism to the epigastrium, in order to prevent what is drunk from being rejected by vomiting.* This form is one of the most fatal, and has caused the greatest consternation in towns and villages, where it has

---

* To the symptoms of this form of the epidemic, as witnessed at Huaras —on the Pacific side of the Andes—Dr. Macedo adds "hiccup" as generally present. *Restaurador de Huaras,* 11th Nov. 1851.

been confounded with the yellow fever, on account of the rapidity with which it snatches away its victims ; and, also, because in some cases the yellow colour is so strongly pronounced that it stains the clothes and bed covers, even after death." [By this observation we learn, that Dr. Daza was not yet persuaded by the startling facts which he relates, to abandon the idea of 'typhus or typhoid' as conventionally agreed, before he took leave of his colleagues in Huaras.]—A. S.

3. *The Nervous or Ataxic Form.* — This variety of the epidemic, Dr. Daza informs us, is more fatal, but, happily much less frequent than the preceding. It was sometimes of only a few hours duration, and therefore required the most prompt assistance.

" *Symptoms.*—It begins with the same signs as the benignant form ; but from the third to the seventh day, a series of alarming symptoms present themselves. These are a tranquil delirium, on other occasions furious raving, and almost always of an intermitting kind. Sometimes the patient falls into a profound sleep—a sort of lethargic state from which he cannot easily recover himself ; the pulse is small and frequent ; and, at times, there is a species of tremor and convulsive motion in the neck and arms, or confined to the hand only. But, in fatal cases, the eyes are turned, or squint ; and the patient's whole frame is seized with tremblings. The tongue is sometimes cold and pale ; the headache is in most cases intolerable ; appetite totally gone ;.thirst very little ; the wasting of the body is very observable in this modification of the disease.* Should the sufferer survive the fourth or fifth day, his convalescence will be extremely protracted."

In the year 1860, Dr. Daza was returned as one of the deputies of Congress in Lima. I then took the liberty of addressing to him a note, requesting some explanation of a few points which appeared to me not quite clearly stated in his *Piscobamba Treatise on the Epidemic Fever of Conchucos,*

I particularly wished to be informed, whether he meant to say, that the skin never assumed the yellow hue in his benignant form, which he described as eminently hæmorrhagic ; and if he intended to exclude hæmorrhage from the symptoms of his bilious form, as he did not mention it in his general description. In a note dated September 28, 1860, Dr. Daza obligingly replied to my inquiries, and stated that,—" In the

---

* Among the more rare or anomalous symptoms of the yellow fever in Cadiz, Dr. Arejula notices the sudden wasting of the body. *La Extenuacion casi Repentina del Enfermo.*

benignant form, the yellow colour of the skin was very little observable ; that in the bilious form nasal hæmorrhage was of very frequent occurrence, as it also was in the other forms of the disease, though of different symptomatic value.   The petechiæ he considered as common to all the modifications of the epidemic fever, and to all its stages, with this only difference, that he met with these in greater number and of larger size, in the two latter of the forms described by him, and particularly in the more advanced period of the disease."

After receipt of this satisfactory and clear statement, I had the pleasure of several personal interviews with this intelligent young physician.   He informed me, among other things, that the appearance of the epidemic, in its various symptoms, was more or less modified by climate, and that he published his account of it from his experience in Vilcabamba and Piscobamba.   But, afterwards he observed that oozing of blood from the gums was a common symptom of the epidemic in the warm district of Siguas, and that he had seen some instances where the gums and throat had sloughed or mortified.   In reference to the different value to be put on nasal hæmorrhage, in the different forms of the fever, Dr. Daza remarked, verbally, that when epistaxis came on early in the case—as it often did in the more simple benignant variety, it was a favourable sign, and often critical ; whereas, when it did not appear until the fourth day or later, it was a bad sign, accompanied with great prostration of vital power, and then the blood was passing rapidly to a state of dissolution.

Earliest Phases of the Epidemic in the Province of Huari. 1853-54.—Dr. George Hall, who attended the sick at Yurma, and Dr. Manuel Santa Maria who assisted in Chacas, being both dead, I applied for information to the curate of the latter place, which is a parish of 12,000 or 14,000 souls, spread over a diversified range of climate.   This reverend gentleman, by name, Don Anselmo Pardo de Figueroa, kindly furnished me with the following particulars:—

1. The individual who introduced the disease into this curacy was a soldier deserter from the coast, and a native of Piscos.

2. The first who conveyed the same disorder to the town of Chacas, was Baltazar Falcon.   This man caught the contagion in Piscos, where he went in search of a dead man's effects, and was carried home on a litter, the distance of six leagues. Those who came in contact with him took the fever, which soon became general in Chacas.

3. The Rev. Don Anselmo Pardo de Figueroa was himself

severely attacked by one of the forms of the epidemic with these symptoms—viz., headache, exhaustion of vital forces, pain in the bowels, a burning heat along the throat and œsophagus, vomiting, purging, and fever, almost all at one and the same time. His convalescence was exceedingly tedious.

4. At the outset of this pestilence in the year 1854, the curate, Figueroa and his two assistants, sometimes administered spiritual aid, in the way of brief confession and absolution, to as many as two hundred souls daily.

5. At the commencement in Chacas, the majority of those attacked died in three or four days ; and, in certain instances, those assailed were carried off within two hours. The attack was sudden.

6. Many died with hæmorrhage, but others were observed to die, even more suddenly, who had no hæmorrhage, and whose bodies stained the ground of a yellow colour.

7. Of those who had emissions of blood from the nose or mouth, and small petechiæ, or red spots like flea-bites on the skin, many escaped. Some others whom the confessors had left as dying, while suffering under the more grave hæmorrhagic form, recovered ; but of those whose bodies were covered with dark or livid spots the size of a "real," or sixpence, *not one recovered.*

8. Yellowness of the skin was a predominant symptom.

The Fever of Huari : its Origin : becomes Epidemic in Yurma and Piscos.—In Dr. Daza's memoir of 12th of September 1855, addressed to the prefect of the department of Ancachs, in which he gives the report of his professional doings and mission in Conchucos, he treats among other things of the " origin of the epidemic," the how, where, and the when of its springing into existence.

He says,—" All the current reports regarding the commencement of the epidemic of Conchucos are perfectly uniform in their details,—at least they are agreed in admitting that a *deserter* was the first who imported the disease. It is generally allowed by the principal people, that in the month of September 1853, a soldier deserter on his way back from Lima, was the first taken ill of this disease on the farm station, called Chilcabamba, in the province of Huari." " *His fever was of a kind unknown in the province. It presented the peculiarity of the whole body becoming of a high (" subida") yellow colour, and accompanied with a profuse emission of blood from the nostrils: a thing never before witnessed in those parts."* We thus learn what was the fundamental type of the epidemic of Conchucos or Ancachs—it was the Hæmorrhagic-icteric.*

Dr. Daza sums up his considerations regarding the origin of the epidemic in these terms:—"What can be said with certainty is, that the inhabitants of Chilcabamba, who had bestowed their hospitality on this deserter, were shortly after attacked by the identical form of fever he had, and in so fatal a manner that the whole family succumbed."† The fever then spread rapidly in the neighbouring villages, so that according to Dr. Daza's memoir referred to, it swept off the earth, in the short period of three months, the poor people of Chilcabamba, Yurma, and Pacarrisca, sparing only about one per cent. of their inhabitants.

[It was from Yurma the epidemic passed to Llumpa and Masqui, on the way to Piscobamba. This then was the malady of which such vague reports had reached Lima in the early part of 1854, and to which I alluded at the close of my article on the Rise and Progress of Yellow Fever in Peru, which was published April 1855, in the *Edin. Med. and Surg. Journ.*, and I take this opportunity to rectify an error of date in the concluding paragraph of that paper—page 204, vol. 82 : for 1844 read 1854.]

Incubation.—It is assumed that the first case of this Sierra fever was not occasioned by the heat, fatigue, and other incidents of the deserter's journey, but that he fled from Lima, already impregnated with the seeds of the disorder.

It may be said, that if this deserter had not left the Capital, the primary cause of the disease would have remained latent

---

* I cannot pass over this account of the origin of this epidemic without reminding my reader of the case of the soldier from Cadiz, who, in September 1804, introduced the yellow fever to the Andalusian town named De los Barrios. We are told that this Spanish soldier, of the regiment De Santiago, took lodgings in the town on the 11th, and died in the night of the 12th, or early on the morning of the 13th. The case was reported to the Junta of Health, as one likely to be contagious. A *post mortem* inspection was made eight hours after his decease. By this time, the corpse was all icteric in a high degree ; and petechial, or black and livid spots disseminated, with large livid patches here and there on the body, with abundant signs of the emission of blood by the mouth. The fever now broke out in the houses nearest to the lodging of the dead soldier, and was very soon found to be a declared yellow fever. See *Observaciones Sobre la Fiebre Amarilla,* by Dr. Tadeo la Fuente. Madrid: 1805.

† This account of Dr. Daza's agrees substantially with that of Don Ambrosio Alegre of Caras, in his letter of 5th Jan. 1859, published in the *Lima Med. Gazette.* Chilcabamba is situated in a warm spot at the foot of Piscos, and about half a league from Yurma. These two latter villages stand on opposite sides of the ravine and river, that here separate the provinces of Huari and Conchucos.

in his system for several months, until revived about Christmas. That very year, 1853, the active producing cause, or constitutive element of this yellow fever lay dormant, it is assumed, in Lima and Callao, from the end of July and beginning of August to the end of December; and early in January 1854 (when the soil and atmosphere became heated with a fervid sun), the disease sprung up anew with increased energy. But, how should the delay of several intervening months be the medium of reviving a dormant cause? Will it be answered? by the awaiting of a favourable sunny season.

If solar heat be the condition required for this kind of reproduction, then, we say, that the deserter at once attained it by his flight into a warm and kindly climate. In Chilcabamba, I am informed by the curate, that the chirrimoya ripens naturally on the tree; in Lima, at the same time of year, it needs stoving to bring this luscious fruit to maturity.

The producing cause of yellow fever eludes our powers of perception. It may be a chemical compound, or a simple element capable of combination in different degrees with other kindred elements. On these topics I make no suggestions, and the reflections offered on the deserter's case, are made on the supposition that the seed of yellow fever, whatever in the abstract that may be, is capable of reproduction from one warm season to another, with the interruption of an intervening winter, just as the seeds of plants rest unproductive until the return of a fitting season.

Peculiarities of the Sierra in respect to Yellow Fever.— In Peru, the seasons on coast and mountains are antagonistic; when the weather is thick and vapoury on the sea-board, it is bright and sunny on the Andes. Between the coast and Sierra there intervenes a dry, arid, and lifeless zone of trap and granite. The difference of temperature in sun and shade, on the jagged and glacial ridges of the Cordillera is very remarkable; according as a person happens to be placed in the one or the other of these positions on the same height, he is subjected to scorching heat or chilling cold. The preponderance of solar heat all the year through on the upper Andine platforms, hardly ever permits these elevated regions to bear the wintry aspect of our European Alps for twenty hours together, because the snow that falls in the afternoon or night speedily dissolves on the following morning. The distance from one degree of temperature to another is not ruled by geographical lines, but by elevation. From the valley beneath glowing in tropical vegetation to the Indian village on the cold pastoral height, that immediately overlooks the waving cane fields, or banana and orange groves, is but a bird's flight of a few minutes. The rapid transitions of the Sierra climates, naturally produced modifications more or less sthenic or asthenic in the epidemic as it moved from one village, town, or valley to another; but whatever characteristic symptom of the primary generic cause as manifested at the outbreak of the disease was lost or counteracted by circumstances in the colder, seems to have been again recovered in the warmer climate. So that, upon the whole, its vitality was never exhausted at any season. It was particularly observed in Cajamarca during the dry season, when the days are sunny, and the nights more or less frosty at high elevations, that the progress of the epidemic appeared to be arrested; but that as soon as the fall of grateful showers presaged the approach of the rainy and warmer months on the Andes, the fever acquired greater strength and facility of propagation.

Temperature needed for Yellow Fever.—It is observed by La Roche, on yellow fever—vol. i, p. 592, that a temperature of 60° Fahr. is necessary for its manifestation, and that it never prevails as an epidemic when the temperature of the summer falls below 65°. We shall find that the epidemic yellow fever of the coast of Peru (where the general annual range of temperature is from 60° to 84°*), when conveyed from Islay to the department of Cuzco, retained its vitality in particular individuals as they climbed into colder and colder regions to traverse the great western Cordillera. But in descending eastward, the fever became epidemic at Sicuani, and extended thence to the city of Cuzco, where, if I am well informed, the temperature of summer rarely if ever reaches 65° Fahr. in the shade. Again, in Cerro Pasco the epidemic still retained—we are told on respectable authority, much of its essential character during the dry months of July, August, and September, with a mean temperature of 44° by day.

Transmutation of Type as observed in Lima, &c.—M. Dupée, a respectable French artisan, had a large establishment by the River Rimac. This situation was malarial and exposed him to frequent attacks of intermittents. About the end of September 1860, when the thermometer indicated from 70° to 72° Fahr. he was seized with an intermittent

* On the sandy coast of Piura, and also of Palpa and Nasca, etc., the summer range is as high as 90°.

From an
Original Sketch,
by
W. Harvey, Lima.
1857

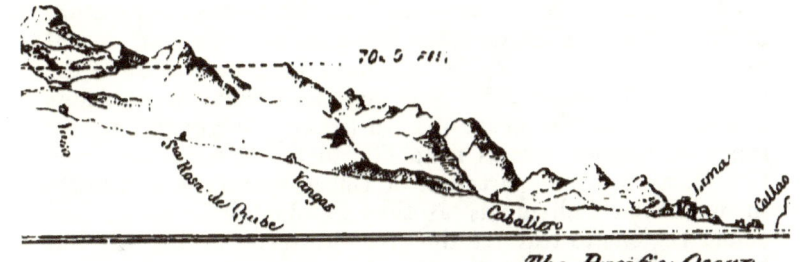

The Pacific Ocean.

the Indian village on the cold pastoral height, that immediately overlooks the waving cane fields, or banana and orange
groves, is but a bird's flight of a few minutes. The rapid
transitions of the Sierra climates, naturally produced modifications more or less sthenic or asthenic in the epidemic as
it moved from one village, town, or valley to another ; but
whatever characteristic symptom of the primary generic cause
as manifested at the outbreak of the disease was lost or
counteracted by circumstances in the colder, seems to have
been again recovered in the warmer climate. So that, upon the
whole, its vitality was never exhausted at any season. It
was particularly observed in Cajamarca during the dry
season, when the days are sunny, and the nights more or
less frosty at high elevations, that the progress of the
epidemic appeared to be arrested ; but that as soon as the
fall of grateful showers presaged the approach of the rainy
and warmer months on the Andes, the fever acquired greater
strength and facility of propagation.

Temperature needed for Yellow Fever.—It is observed by
La Roche, on yellow fever—vol. i, p. 592, that a temperature
of 60° Fahr. is necessary for its manifestation, and that it
never prevails as an epidemic when the temperature of the
summer falls below 65°. We shall find that the epidemic
yellow fever of the coast of Peru (where the general annual
range of temperature is from 60° to 84°*), when conveyed
from Islay to the department of Cuzco, retained its vitality
in particular individuals as they climbed into colder and
colder regions to traverse the great western Cordillera. But
in descending eastward, the fever became epidemic at
Sicuani, and extended thence to the city of Cuzco, where, if
I am well informed, the temperature of summer rarely if
ever reaches 65° Fahr. in the shade. Again, in Cerro Pasco
the epidemic still retained—we are told on respectable authority, much of its essential character during the dry
months of July, August, and September, with a mean temperature of 44° by day.

Transmutation of Type as observed in Lima, &c.—M.
Dupée, a respectable French artisan, had a large establishment by the River Rimac. This situation was malarial and
exposed him to frequent attacks of intermittents. About
the end of September 1860, when the thermometer indicated
from 70° to 72° Fahr. he was seized with an intermittent

---

* On the sandy coast of Piura, and also of Palpa and Nasca, etc., the summer range is as high as 90°.

Section of the Andes of Peru from the Pacific to the Montaña

N.B. Under the dotted line was written *palti ye* the Western Peruvian Andes.

the Indian village on the cold pastoral height, that immediately overlooks the waving cane fields, or banana and orange
groves, is but a bird's flight of a few minutes. The rapid
transitions of the Sierra climates, naturally produced modifications more or less sthenic or asthenic in the epidemic as
it moved from one village, town, or valley to another ; but
whatever characteristic symptom of the primary generic cause
as manifested at the outbreak of the disease was lost or
counteracted by circumstances in the colder, seems to have
been again recovered in the warmer climate. So that, upon the
whole, its vitality was never exhausted at any season. It
was particularly observed in Cajamarca during the dry
season, when the days are sunny, and the nights more or
less frosty at high elevations, that the progress of the
epidemic appeared to be arrested ; but that as soon as the
fall of grateful showers presaged the approach of the rainy
and warmer months on the Andes, the fever acquired greater
strength and facility of propagation.

Temperature needed for Yellow Fever.—It is observed by
La Roche, on yellow fever—vol. i, p. 592, that a temperature
of 60° Fahr. is necessary for its manifestation, and that it
never prevails as an epidemic when the temperature of the
summer falls below 65°. We shall find that the epidemic
yellow fever of the coast of Peru (where the general annual
range of temperature is from 60° to 84°*), when conveyed
from Islay to the department of Cuzco, retained its vitality
in particular individuals as they climbed into colder and
colder regions to traverse the great western Cordillera. But
in descending eastward, the fever became epidemic at
Sicuani, and extended thence to the city of Cuzco, where, if
I am well informed, the temperature of summer rarely if
ever reaches 65° Fahr. in the shade. Again, in Cerro Pasco
the epidemic still retained—we are told on respectable authority, much of its essential character during the dry
months of July, August, and September, with a mean temperature of 44° by day.

Transmutation of Type as observed in Lima, &c.—M.
Dupée, a respectable French artisan, had a large establishment by the River Rimac. This situation was malarial and
exposed him to frequent attacks of intermittents. About
the end of September 1860, when the thermometer indicated
from 70° to 72° Fahr. he was seized with an intermittent

---

* On the sandy coast of Piura, and also of Palpa and Nasca, etc., the summer range is as high as 90°.

From an
Original Sketch,
by
W. Harvey, Lima,
1857

The Pacific Ocean.

fever. He had afterwards several returns of the ague par-
oxysm in all its stages for three or four successive days. On
the 2nd of October, the sixth day from the first attack, blood
was observed to ooze from the gums; minute petechiæ
appeared on the neck and body, as well as on the lower
extremities; but so small as to be scarcely perceptible. The
heat of skin and frequency of pulse were, also, rapidly falling.
The patient had been ill several days before he thought it
necessary to apply for medical advice. When I was called
to his assistance, he was suffering from an ague fit; but
next day when I observed the signs enumerated of a well
marked transition state, I administered a turpentine emulsion in convenient and repeated doses. On the following
day I found that the warmth of the surface had been retained without the least return of fever, and that the
sanguineous exudation from the gums had ceased. The
patient, in the interval, passed two dark and sanguineous
motions. He seemed easier, and continued so until 2 P.M.,
when he was surprised by an attack of epistaxis and vomiting
of dark blood. The petechiæ came out freely, and some of
them were the size of a silver three-penny piece (quartillo)
on the chest, abdomen, and extremities. The surface of the
body assumed a yellow colour, with ecchymosed stripes and
blotches about the neck and shoulders. The case was now
a declared yellow fever. I requested a consultation, and was
met by Dr. D. Mariano Macedo. As to the nature of the
disease there was no room for difference of opinion. We
happily saved M. Dupée's life by the vigorous application of
Dr. Copland's curative treatment, as laid down in his *Dict.
of Pract. Med.* in the hæmorrhagic form of this dire malady.*

When we come to treat of the Sierra epidemic of 1855-56,
in the valley of Abancay, we shall have another occasion to
mention this subject of transition of type. Whether these
transitions be admitted as modifications proceeding from the
same general cause acting under special circumstances of the

---

* This method of cure was found so valuable, and saved so many lives in Peru,
during the yellow fever reign of terror, that I shall here place on record the first
case in which its application was propounded in the capital. It was the fatal
case of Senor Loyo (July 1853), a gentleman from the cold elevation of Huan-
cavilica, who had only arrived a few weeks previously from the Sierra. In its
source it was supposed to have been of the nature of a remittent or intermittent fever, to which persons from the Sierra are peculiarly prone on coming
to the coast in the autumnal season. But, on the first appearance of black
vomit and yellow skin, etc., a medical consultation was called. It consisted,
at first sitting, of six physicians. I proposed to the junta the administration
of turpentine by the mouth to stop the black vomit. I was only supported by
one vote. The remedy was rejected as an innovation. At a later season, its
efficacy was recognised by all.

individual recipient, or as the amalgamated effects of distinct co-existent causes, are nice questions which I shall leave others to decide.

The Epidemic of the Sierra not a Disease of Race.—It has been a vain invention to call this epidemic of the Sierra a disease peculiar to the Indian race. It is very true, that on the Andine regions there are villages, towns, and entire districts of an unmixed Indian population. In such places and circumstances only Indians could be the victims; but the Rev. D. Anselmo Pardo de Figueroa, curate of Chacas, had some experimental knowledge of this subject, in his own person, though of Spanish descent. He says—"The disease was not of the Indian people, as too carelessly assumed, but rather of condition and circumstances." He explains: that under more favourable circumstances, the poor Indian might have got off as well as the better accommodated "*Dons*" of the same indigenous race, who had more means, and lived in houses instead of huts—in abodes of sufficient ventilation, and fitted with compartments, in which the sick, if necessary, could be kept apart from the sound. The fact seems to be established on the widest scale of comparison, from one end of the Republic to the other, that no rank, class, or race, were certain of not being attacked by this pestilence; while it is admitted that the poor, badly-clothed and fed, neglectful of cleanliness, and congregated with their guinea-pigs, in low, dark, unventilated hovels, without chimney or window, sleeping in their scanty clothing on sheep skins, or huddled on the floor of one confined and filthy apartment, were, by the very misery of their condition, doomed to suffer the great burden of this general calamity. I may add that the negro race, who were comparatively free from yellow fever on the coast, do not inhabit the Sierra, with some rare exceptions, on the sugar estates.

The Epidemic Contagious.—On this important subject, Dr. Daza, in his *Memoir* of the 12th September, 1855, already referred to, observes—That from Chilcabamba and Yurma, as a focus or centre, the epidemic radiated in different directions. "Not with that rapid march which characterises diseases of infection, but, on the contrary, with the slow, and up to a certain degree stationary, progress of a contagious disease; remaining in one place or town, three, four, and even six months; moving onward in a successive manner, from one station to another, without leaving an intermediate farm or hamlet free from the mortiferous visitation."

The Mortality of the Disease greatly controlled by "Hygeia."—Dr. Daza, under the heading "General Dietetic Observations," assures us that the greater number of the native Indians who perished in Conchucos during the reign of this fatal scourge, were, in reality, not carried off by the disease. He ascribes the great majority of deaths to individual circumstances, and particularly to disorders in diet in time of convalescence, during which the appetite was at once craving and insatiable. He then goes on to show the practical effects of Hygiene, and medical treatment in the well-ventilated lazarettos under his care, for the particulars of which we must turn, once more, to his memoir of 12th September, 1855, addressed to the Prefect of Ancachs.*

Lazarettos.—Dr. Daza tells us, that when he entered the district of Piscobamba, in the month of October 1854, he found the local authorities were so engrossed with political affairs that they showed entire indifference to the people. Thus, unregarded, the Indians struck by the epidemic, dropped by hundreds, and became food for the birds of prey. Under these cheerless and heart-rending circumstances he opened his first lazaretto on the agricultural estate called Vilcabamba. In this place of refuge above one hundred and eighty sick were treated during the months of November and December, and all cured, notwithstanding there were among them some cases of much gravity.

After this, in the months of January and February, and up to the middle of March 1855, we find Dr. Daza in Piscobamba, attending a lazaretto there, in which he was almost equally successful as in Vilcabamba. Out of two hundred of all ages and sexes admitted, all were cured except two.

Siguas, or Sihuas, the Capital of Conchucos.—Topography.—This provincial city is situated in a narrow glen, bounded on the north-east and south-west by lofty and almost perpendicular mountains; while towards the north-west and south-east it is open to ventilation. The locality enjoys a dry and hot climate, in which intermittent fever is not endemic. Its soil is argilaceous, and produces delicious fruit as well as plenty of corn. In this favoured place Dr. Daza opened a lazaretto, on the 15th March 1855, with eighty individuals sick of the epidemic fever. As these were gradually dismissed, completely cured, the numbers that

---

* For the perusal of this interesting document, I have to acknowledge my obligations to my most obliging and highly gifted colleague, Dr. Jose C. Ulloa of Lima.

afterwards entered the asylum increased progressively; so that in the three months April, May, and June, upwards of six hundred persons of all ages and sexes received medical treatment, and all recovered except eight individuals. In these fatal cases—which presented themselves in a grave state—the epidemic was complicated with other most doleful accidents, such as abortion, gangrene, consumption, and œdema of the feet, &c.

Average Duration of the Epidemic Fever in the Lazaretto of Siguas.—Taking the data furnished by Dr. Daza, of eighty persons in the latter half of the month of March, and upwards of six hundred treated in the three following months, we have a total of about seven hundred individuals of whom nearly all were cured in one hundred and seven days. If we divide the sick by the number of days, we shall arrive at the same result which, as we shall see by and bye, Dr. Don Pedro Espinosa did in his lazaretto of Ayacucho: that is, that the disease, as a general rule, was cured before the eighth day, when it ended favourably. In Huaras, however, according to Dr. Macedo, the disease was of much longer duration: probably owing to the greater cold of that locality.

Distinctive Character of the Epidemic of Huaras.—From the Restaurador of Huaras, of 11th November 1854, and from his paper on the Sierra epidemic, read before the Medical Society of Lima, 15th February 1856, we learn that Dr. D. Mariano Macedo, had no opportunity of seeing the hæmorrhagic-icteric variety of this disease. He distinguishes the hæmorrhagic form into the " simple," and more " severe," without yellow colour of the skin We have, therefore, the peculiarity, as observed by this physician, that yellow skin only accompanied the non-hæmorrhagic form, which he calls the hepatic, in Huaras.

On other points, however, Dr. Macedo's general description of this epidemic, is in accordance with Dr. Daza's. He tells us that in Huaras the fever underwent different modifications, after the close of the first stage.

Generally, from the fourth to the seventh day, when the second stage had set in, the disease assumed different forms, as the hæmorrhagic, the hepatic, and the cerebral.

The symptoms of the first stage were common to all the forms of the fever: whether the simple, more severe, or fatal.

In the second stage, therefore, there sprung up such defined differential signs as characterised the distinct forms enumerated.

Dr. Macedo remarks, that, in Huaras, the hepatic was less prevalent than the hæmorrhagic modification, but that it was the more severe and fatal. When the disease assumed this particular aspect, he observes: "The skin began to show a yellow colour, which was of a saffron hue in some, as in the most strongly marked icteric case of yellow fever." It will be observed that this corresponds to the "bilious" variety of Dr. Daza.

The Epidemic of Yurma transmitted to Caras and Yungay. —From Yurma to Huaras, Yungay, and Caras, there are distinct tracks across the breaks of the Great Western Cordillera. Two of the highest farms on the Pacific side, are Santa Cruz and Vicos, rented at the time of the pestilence by Don Ambrosio Alegre, a well-known sugar planter in Caras. To these lofty farms the fugitives from Yurma early directed their steps, but the evil which they thus hoped to escape, already circulated in their blood. The malady spread in their line of pilgrimage, and became general in the province of Huaras. Strong young people who were seen at work in the fields in apparent health, were, sometimes, so suddenly struck powerless by the epidemic poison, that in a few hours after, they were stretched out lifeless, with their bodies of an intense yellow colour; and, according to Mr. Alegre, they stained the ground yellow underneath them. This gentleman kindly favoured me with two letters, one of the 5th January 1859, and another of the 29th July 1860, on the subject of this epidemic.[*]

On his sugar estate Cayñasbamba, situated in an ardent climate, he established, at his private expense, a lazaretto in which there were from eighty to ninety patients at a time, from September to December 1854. It was observed by Don Ambrosio Alegre, during his constant superintendence at this lazaretto, that, generally on the third or fourth day of the fever, there appeared on the skin red or mulberry coloured spots of different dimensions: the latter, when large, being of the worst omen. The body assumed a golden

* In the first of these letters, which was published in the *Lima Medical Gazette*, it was stated that, in 1856-57, when General Vivanco's encampment was at Casma (the seaport of Huaras), cases of yellow fever and black vomit were conveyed by infected persons—"contagiados"; but that, instead of this vomit, the fatal symptoms, among those affected in the Sierra, were the dark maculæ, and red spots in the skin. It would, therefore, seem to be a fair inference from this fact, that in the climate of Caras and Yungay, to which the observation referred, at the supposed elevation of from six to seven thousand feet, the blood poison of yellow fever, imported from Casma, directed itself chiefly to the cutaneous capillaries in preference to those of the stomach.

huc, and there were many cases attended with hemorrhage.
The medical treatment in this lazaretto was limited to very
few drugs, viz., tartar emetic and Epsom salts, at the commencement of the attack ; iced lemonades at the height of
the fever ; and in the latter stages quinine. In fact, after
clearing out the first passages, the treatment was principally
hygienic, and out of three hundred and fifty individuals on
Cayñasbamba, more than two-thirds were seized with the
fever, of whom only three men and six women died. On the
more distant farms, however, where he could not render
early assistance, the mortality was great.

The town of Caras contained three thousand inhabitants,
nearly all Indians, yet only three or four deaths occurred
among them from the epidemic. But in the annexed rural
district of Yañaguara, out of six thousand of both sexes and
all ages, no fewer than three thousand died of this pestilence.
The contrast is prodigious. Mr. Alegre, however, ascribes
the difference in result to want of accommodation, cleanliness,
and the general condition of life, among the rural Indians
of Yañaguara. See *Note B.*

*The Epidemic of Conchucos spread beyond the bounda-
ries of Ancachs.*—Leaving the fertile districts of Yungay
and Caras, we now turn towards the Eastern Cordillera
range, and those richly varied and fruitful regions watered
by the Huallaga and its tributaries.

We have seen how this pestilence was imported from
Piscos to the town of Chacas, and from Yurma to Iluaras.
From these points it spread southward to St. Luis, St.
Marcos, Chavin, and Cajatambo. On the other hand in the
village of Yanama, nearly opposite to Yurma, out of one
hundred individuals attacked in 1854, only five escaped with
life. Pursuing a northern course it overran the districts of
Piscobamba, Siguas, and Huamachuco. In the province of
Pataz it so completely destroyed the population of Chongos,
that the huts were thrown in upon the dead to serve them
as graves. The epidemic still pressed on its way in the
direction of the Marañon. In Tayabamba it proved most
fatal, and did not spare Huaylillos or the capital of Pataz,
Condamarca, and Cajamarquilla : the latter being on the
confines of three adjoining departments, La Libertad, Amazonas, and Cajamarca : not one of which escaped from its
successive attacks. Its march was like that of the locusts,
that leave one tree bare of its leaves before they fall upon
the next. Chota suffered excessively, notwithstanding the
efforts made in 1855 by the Prefect of Cajamarca, to prevent

the spread of contagion from Pataz.* From this latter province it slowly reached the remote city of Chachapoyas, the capital of Amazonas, in south latitue 6° 15'. In Moyobamba, the capital of Loreto, the epidemic assumed a most peculiar and fatal form. The Prefect and Bishop of Chachapoyas, in October 1857, sent officially to the supreme government in Lima, a most calamitous report of the mortality of the pestilence as developed in those parts.

Fatal Epidemic of Moyobamba.—The Bishop of Chachapoyas and his coadjutors describe the disease as having fallen heavily on the lungs, causing great disturbance of the respiratory organs, with an intolerable fetor of the breath. This latter symptom probably indicated a process of pulmonary gangrene, like some of those malignant forms with pulmonary complications, pointed out by Arejula in his autopsies. Shortly before death there was a great change observed in the colour of the skin, for the body assumed a dingy hue. This ominous sign may be looked upon as evidence of the rapidly putrescent tendency of the disease, the duration of which did not exceed three days at the most.†

The Epidemic of Conchucos transmitted through Huamalies to Huanuco.—The valley of Huanuco is a little world by itself, including almost every gradation of climate. On the east it is bounded by the E. Cordillera, whose craggy pinnacles are only frequented by the soaring condor, and on the west by the lower and central Cordillera chain, which separates it from Huamalies and the valley of the Marañon. It is watered by the upper Huallaga and many mountain rivulets. The eastern side is clothed in natural verdure towards Ambo, but at the bottom and on the western side, arid (only producing tufts of cacti) on the hill sides during eight months in the year. Here it rains little—the annual range of temperature during three years that I resided in this place did not vary 10° Fahr.; and altogether the temperature is as equable as its soil is unequalled in fertility. This valley is the great portal to the " Montaña," or wooded

* Paper entitled " Razon de las Providencias del Gobierno", or an Account of the Measures taken by the Government to Combat the Epidemic, in the Archives of the Minister of Justice, Instruction, and Beneficence, etc.

† See my contribution, entitled " Geography of Diseases in the Climates of Peru", Jan. 1858, *Edin. Phil. Journ.*, new ser., vol. vii. I may take the opportunity to notice that what is said in the article now referred to, on the German *importation* of yellow fever into Peru in its nascent form of 1851-3, is partly neutralised and further explained by the more recent information embodied in the introduction to the Sierra epidemic of 1853 7. See *Note*, p. 285.

territory of boundless vegetable riches, on the eastern side of the Andes.   It is entirely free of malaria, while it is uniformly refreshed by a daily breeze following the course of the river through the break of the eastern barrier. A mountain range of 2,500 feet separates it from the valley of Chinchao, famous for its cinchona and cocoa, which is the great staple of commerce with Cerro Pasco. To the latter place from the bottom of the valley of Chinchao, on the river Huallaga, there is I think an elevation of more than ten thousand feet. Over the whole of this district the epidemic extended from the valley of Huanuco, where it first appeared in the way related in the following abridged letter from one of the principal gentlemen and landed proprietors of the district:—

" Huanuco, 12th July, 1860.

" My DEAR SIR,—I mean in this letter to give you some account of the yellow fever which desolated this province in the year 1855.   This visitation came to us from the coast by the way of Ancachs, committing the greatest havoc in its transit among all the farms and villages.   We contemplated its approach with anxiety for months, till at length it entered this province by the valley of the Igueras.   The first victim of this odious pestilence in the valley of Hongoimaran, where I have my sugar estate, called Vichaycoto, two leagues to the south of the city of Huanuco, was a workman of my own, called Domingo Falcon, of the indigenous race. He caught the infection from an individual who came from Nausa and Chaulan, villages in a cold climate on the way to Huamalies. Falcon did not long survive the attack. His case was ushered in by a most violent headache and intense fever, which was attended by hæmorrhage from nose and mouth, over which ice had no power. He died twelve hours from the invasion, on Candlemas day, 2nd February, 1855, and after death his body became all over of a yellow colour. I had as a nurse in my own family—a robust young girl of twenty-four years of age, named Agueda, who was all of a sudden seized very violently. She was struck with intense headache, bodily inaction and fever, almost at one and the same moment ; her skin became so hot that it felt to the touch like a heated iron. She became insensible, and raved in her mind. In less than twenty hours from the invasion she was dead, and her body became of a livid (and yellow) hue.

" The last who died in the valley was (an Englishman) Mr. John Dyer of Quicacan. The fever attacked his head, and caused delirium with insensibility. The skin and tongue had dark stains or spots, but without hæmorrhage from the nose or mouth, which usually accompanied this petechial form in Huanuco. Mr. Dyer died on the eleventh day of the fever, and was carried off in strong convulsions.

I remain your affectionate friend,

MIGUEL YNGUNZA.

To Dr. Archibald Smith, Lima.

Don Miguel Yngunza, also furnished me with further particulars in a separate document relative to this epidemic, from which I shall make a short abstract to avoid unnecessary repetitions.

At the first outbreak of the epidemic at Vichaycoto, where he resided at the time, he had upwards of one hundred individuals on the property, of whom, as the disease progressed, he had as many as thirty, or more, in his lazaretto at a time. He himself acted the part of hospital assistant and mediciner, until towards the close of the epidemic, when he had the good fortune to secure the professional services of Dr. G. A. Mulgrew.

General features of the disease in Huanuco.*

1. Mulberry-coloured spots on the skin were almost general in the Indian, but less constant in the white race.

2. In the more severe cases, livid or black spots appeared on the tongue; blood very commonly exuded from the gums, which became swollen, and when the disorder extended to the throat, swallowing became difficult or impossible.

3. In a great majority the skin took on the yellow colour of the flower of the broom, either during life or after death.

4. Sometimes the urine was suppressed, and the bowels so much confined that they could not be relieved in the ordinary way by glysters.

5. There were cases in which those attacked by the epidemic lost their senses; became violent, and beat or tore themselves furiously. Their nails and lips used to turn blue before death, which took place in two or three days.

6. Many affected with petechiæ of a dark colour, were carried off on undeterminate days, during the first week of the malady, with profuse nasal hæmorrhage, or with vomiting of very dark blood.

Besides these special modifications of the Huanuco epidemic, Mr. Yngunza supplies the following general remarks.

While the epidemic spared no age or sex, it was always

---

* The city of Huanuco is by barometrical measurement, according to Rivero, 1,943 metres above the level of the sea.

observed that the young and robust were more violently assailed than the more frail and less vigorous.

The sure sign of amendment was the cessation of the headache. He that perspired freely recovered, and those who survived the tenth or eleventh day, were usually considered out of danger. But there were cases of longer duration with constant fever.

After convalescence had commenced, relapses were frequent and easily produced, especially by errors in diet; they were not few who died in this way.

N.B.—*A Curious Episode.* After the decease of Mr. Dyer of Quicacan, on 7th of September, 1855, the pestilence disappeared in the valley of Huanuco. Early in 1856, Mr. and Mrs. Yngunza made a visit to Lima, where, at this season, yellow fever reigned, especially among strangers. A favourite man servant of Mr. Yngunza's caught this fever in the capital. The man recovered after a dangerous illness, and when able to travel accompanied his master back to Huanuco—a tedious journey of seventy-four leagues. In eight days they arrived on their estate of Vichaycoto, where there had been no sickness for several months. But just four days after their arrival at home, Mr. and Mrs. Yngunza were simultaneously seized with fever, the same as they had seen in Lima, and previously in their own lazaretto. The skin in the lady's case assumed the yellow colour. In both husband and wife blood exuded from the gums, and in the gentleman's case this symptom was exceedingly severe, so as to have affected his teeth. Towards the close of the fever, which lasted eleven days, he had evacuations of blood. Convalescence was very tardy. Now, a question arises :—Was it the yellow fever of the capital or that of Huanuco revived, that attacked Mr. and Mrs. Yngunza? From them the disease was not propagated to others. The valley remained healthy.

Valley of Chinchao.—From Huanuco the epidemic penetrated to Chinchao, the adjacent "Montaña," where most of the wealthy proprietors of Huanuco cultivate the Erythroxylon coca. This is a warm and humid district, on the eastern slope of that range of mountains which shut in the bottom of the valley of Huanuco, from those regions which look to the great basin of the Amazon. Here it was expected that the disease would have assumed its most fatal character. But instead of that, it was completely disarmed of danger by the empirical treatment adopted by the Indians on the spot, as we shall notice under the head of "Treatment."

Cerro-Pasco.—This high mining town (twenty-two leagues from the city of Huanuco) is the capital of the rich and varied department of Junin, which includes Huanuco among its provinces. It is situated just under the ridge of the eastern Cordillera at an elevation of above 14,000 feet in south latitude 10° 26′. Population fluctuating, but estimated at 18,000, at the time of the epidemic. It was introduced here from the Huanuco side. It overran the districts of Yanacancha and Mataderia, chiefly inhabited by the poorer population. The centre of the town, occupied by the wealthier people, suffered comparatively little. Five or six leagues lower, towards Huanuco, by the Huacar route, we have the curacy of Chacayan, in a temperate grain growing district. In March 1855, the epidemic arrived here. It attained its height in May, and nearly disappeared by the end of August. The Rev. D. M. Saria, curate of this place, informed me of the following interesting particulars. In the lower portion of the parish of Chacayan, there is a village called Atapilca. Its population consisted before the epidemic of one hundred families, of about five or six individuals in each. They all lived in crowded, dark, and miserable huts, so, that during this horrid pestilence they were reduced to the necessity of enduring the foul air of their own bodies. Of all the wretched inmates of Atapilca not more than thirty survived the ravages of the disease. Out of five thousand parishioners, he confessed and buried one thousand three hundred. The epidemic carried off all the young and able-bodied of both sexes.

At Junin the Epidemic of Conchucos terminates.—From Cerro Pasco, the epidemic extended to Nina-Caca, whence it diverged eastward by Huancabamba to the " Montaña" of Paucartambo, on the confines of endless forests ; and southward it continued its progress along the eastern side of the lake Chinchaycocha, to the village called Carguamayo. Here it seems to have come to a stand, for the town of Junin, situated in a plain by the southern extremity of this lake, we are told was not touched by it. At this point, then, I should draw the boundary line, or southern limit of the epidemic of Conchucos.

The central Andine pestilence which broke out in Chincheros, appears to have radiated northward, as far as Tarma,*

---

* On the Jauja route of the battalion Aymaraes, some stray cases are reported to have sprung up at Chacapalca and Apata, and may possibly have kindled the conflagration in the neighbouring districts. Dr. Espinosa ex-

and southward to the borders of the Apurimac, where it met
and mingled with the great Cuzco epidemic, which commenced, as we shall explain in Sicuani.

Remark.—I have now brought my review of the different
phases of the fever of Conchucos to a close. In tracing this
epidemic through different regions, I have endeavoured to
base all the most substantial and fundamental descriptions
of the disease on precise medical evidence, and to illustrate
and confirm the same, by adducing the experience of intelligent non-medical gentlemen, such as the confessor of the
sick, and the founders of lazarettos—persons who had themselves direct opportunities and strong inducement to attend
to the facts of which they took cognisance. All these eye-
witnesses, whether medical or non-professional, testify to the
same general facts, while they mutually supply each other's
deficiencies in particular details, peculiar to different temperatures and localities. By this wide foundation of *data*, it
is to be hoped, that partial judgments on too limited a scale
of observation may be avoided, and that the reader may
arrive at just conclusions from the whole evidence now laid
before him. (See *Note C.*)

PART II.—THE EPIDEMIC OF AYACUCHO.

Having entered so fully into the details of the epidemic
of Conchucos, I shall avoid any unnecessary repetition of
similar symptoms in tracing the origin and general features of
the same disease as it was observed in central and south Peru.

On the 22nd February 1855, General San Roman, then
the Minister of War, issued a decree granting to all the
citizen-soldiers from the southern departments permission to
return to their respective homes.

The battalion Aymaraes, with destination to the inland
province of the same name, in the department of Cuzco,
took their homeward route through Jauja, Ayacucho and
Andaguaylas. The troops left the capital in apparently a
healthy state, and were about to traverse transandine districts also free of any known general disease. Lima that
year (though infested with intermittents) was free of epidemic
yellow fever, but it prevailed in Trujillo among the battalion
Cuzco.*

---

pressly states, that in no instance did the epidemic of Ancachs appear in the
army of General Echenique, while in the province of Jauja in 1854. *Lima
Med. Gazette*, 15th July, 1859.

* See Col. Narciseo Arestequi's letter to Dr. Garvi:o, in the *Resena
Historica.*

Don Mariano Herencia Zeballos, Colonel of the battalion Aymaraes, took leave through the *Comercio* newspaper, of his friends in Lima, on the 24th March 1855, and on the 5th of April he entered the city of Ayacucho, at the head of about three hundred men. On this hurried march over an estimated distance of one hundred and ten or one hundred and twenty Spanish leagues, totally unaided by commissariat arrangement, there was naturally desertion and indisposition on the way. On his arrival at Ayacucho, Col. Zeballos was soon provided by the intendant, Don Tadeo Duarte, with means of transport for thirty or forty sick and weary whom he did not like to leave behind. By the evening of Easter Sunday (8th April) the soldiers of the Aymaraes with their female followers, were in march southward by Ocros, Pampas, and Bombon. They had, however, scarcely emerged from the hot mosquito territory, called the low grounds or Bajos of Bombon, when the first death of yellow fever was seen among them.

In a letter of the 20th July 1855, from Don Paulino Guillen, the governor of Chincheros, addressed to the titular physician of Ayacucho, Dr. Lucas Alejandro de Murga, we find it stated that, "the yellow fever of Chincheros originated in the case of a soldier who died in Bombon on a certain day in April; and in the month of May twenty-five licensed soldiers with six 'rabonas' (female camp-followers) succumbed to the same disease. The sick and the dead exhibit a yellow colour on the lips, teeth, nails, and all the body." In another letter to the same physician, of 23rd July, the Rev. Aniceto Jeri, acting curate of Chincheros, says, that the mortality in his parish is excessive on account of the "yellow fever," and that so many die on the hills that their numbers are incalculable. The infected corpses are viewed with horror and cannot be conveyed regularly to the parish burying ground.* From Chincheros this fatal malady spread to Cocharcas and other neighbouring towns of less note. The great fair of Cocharcas in September brought crowds of people together, and as these dispersed and returned to their homes in Cangallo, Ayacucho, and other parts, they carried with them infection and propagated the fever.

The Epidemic of Chincheros and Cocharcas transmitted

---

* I have taken these extracts from the authentic documents of a meeting held at Chincheros relative to "la fiebre amarilla" or yellow fever, under the sanction of the public authorities; for the perusal of which I was indebted to my obliging colleague Dr. Jose C. Ulloa.

to Ayacucho.—Dr. Pedro Espinosa was appointed to the charge of the lazaretto of Quicapata, in the immediate neighbourhood of the city Ayacucho, and has favoured me with two letters on the subject of the epidemic. One of these letters, dated the 16th April 1859, was published in the Lima *Medical Gazette,* and from the other of 29th July of the same year, now before me, I make the following extract:—" Generally, from the fifth to the sixth day, there was profuse epistaxis, accompanied with maculæ, or large blackish spots on all the trunk of the body. At this stage of the disease the headache, which had been constant and intense from the beginning of the attack, ceased. About the same time the skin assumed the yellow hue of the broom-blossom, as did, likewise, the conjunctivæ, which at the commencement were injected with blood, and in the progress of the fever took on the yellow tint in common with the skin."

Dr. Espinosa further draws the following distinction between the epidemic yellow fever as witnessed by him in the temperate climate of Ayacucho, and the ordinary tabardillo of the Sierra.

From the invasion to the close of the epidemic fever, it was attended with a languid or drowsy state of the perceptive faculties, a vacillating gait, and great prostration of vital forces. It terminated favourably by perspiration (which happily was the more usual result) or fatally by hæmorrhage: rarely, before the fifth or sixth, but, generally, before the ninth day.

In the Sierra *endemic tabardillo,* the commencement of the fever is always sthenic. The primary ardent or inflammatory period is barely concluded by the end of the first week, and then the succeeding stage with its adynamic and ataxic symptoms—[such as a small and frequent pulse, dry tongue, and fuliginous teeth, arid skin and petechiæ, delirium, &c.]—does not terminate before the fourteenth, or, it may be, the twenty-first day. In the tabardillo, hæmorrhagic attacks are not frequent, and when epistaxis occurs it is for the most part of good omen, or critical. The petechiæ (not unfrequent) in the tabardillo are very small, and of a higher colour than in the epidemic fever—besides, they do not appear before the end of the first week. In the epidemic convalescence was very slow; but in the endemic it is more easy and rapid, and less liable to relapses. Tabardillo is not essentially contagious, whereas the epidemic was eminently contagious.

To prevent overcrowding, and guard against relapses, Dr.

Espinosa had a separate sanitarium for convalescents at the distance of a quarter of a league from the lazaretto of Quica-pata. To this provision he ascribed much of his success in the cure of the epidemic. (See *Note D.*)

The Epidemic of Chincheros extends to Abancay.—In the valley Abancay, and at the distance of three leagues from the provincial capital of this name, is situated the extensive estate called Auquibamba, lately rented by (or the property of) Dr. Teodoro Ureta, who resided upon it when the epidemic of 1855-56 reigned there.* This gentleman, by profession a physician, but independent of medical practice, favoured me with the following information.

The boundaries of Auquibamba, stretch from the river Pachachaca to the cold and lofty pasture lands, and consequently the estate partakes of great diversity of atmospheric temperature. On the low grounds the summer heat indicated by the thermometer of Reaumur is 24° and the winter temperature 17°.

The sugar plantations are subject to malaria, and during the epidemic of 1855-56 many cases of this disease terminated in the remittent or intermittent form, when they readily yielded to the free use of wine and bark. Other cases were observed during the same period, which began as intermittents and ended in the continued epidemic form.

The constant and unfailing symptom of the epidemic was pain in the head and eyeballs, which latter were red and injected at the commencement, when the patient could not look up without great uneasiness in the orbits. The skin became yellow, as did also the white of the eye, in course of the fever. Very many had epistaxis, others had maculæ or petechiæ, and vomiting or evacuations of blood. In protracted cases new complications arose, and when vital action was at a low ebb, sores like issues opened on the skin.

Dr. Ureta was peculiarly struck by the pathological physiognomy of the sick. Almost all of them had a remarkably dull heavy look, some with pale and others with flushed countenances; some in a state of perfectly helpless prostration, and others with high fever and pulse—one requiring cordials of wine, camphor, &c., while others, especially on the colder heights, actually needed the lancet.

The Rev. Dr. D. Angel Cardinas, of the large Indian

---

* Close to Auquibamba, the remains of the battalion Aymaraes, thinned by disease and descrtion, turned off the Cuzco line of road to their own province.

town Huancarama, escaped from a severe cerebral and congestive form of this fatal pestilence, by the aid of cupping and blistering. Out of two hundred attacked on his own property, Dr Ureta, in spite of his best efforts to save all, lost thirty individuals. He found that when early assisted by a simple emetic or purgative, the mortality was greatly diminished. In conclusion, Dr. Teodoro Ureta, in a memorandum of 9th April 1860, addressed to me on the subject, sums up the character of the epidemic of Abancay in the following terms:—"The most prominent symptoms were headache and pain in the eyes; excessive heat and thirst; a yellow colour of the whole body; nasal hæmorrhage, and a total prostration of strength."

PART III.—ON THE YELLOW FEVER IN CUZCO.

In the *Comercio* newspaper, of 25th April 1855, we peruse a communication from Islay, of date 21st April, intimating that the town was being deserted :—" To escape from the yellow fever the terrified inhabitants have emigrated to Arequipa, Tacna, Vitor, or Camana, &c." (See *Note E.*) The writer goes on to give a thrilling description of the fatal forms of the malady ;—in some it was signalized by bilious vomiting at the invasion, but soon ending in blood, and proving fatal within twelve hours :—others again, succumbed under evacuations—first of coagulated, but ultimately of " living" blood—within three days, as did also those attacked in the " lethargic" form. The writer then asks, What are we to do ? and answers, " *Esperar la muerte*," wait for death. Such then was the condition of the sea-port of Islay, just before the arrival of the battalion Cuzco, which left the capital on the 15th May. The colonel of this division of licensed troops was D. Narcisco Arestequi, who in a letter of 5th July 1856, addressed to the medical commissioner Dr. C. Garviso in Cuzco, and published in the *Resena Historica*, of the epidemic by the latter, informs us —That many soldiers of the battalion Cuzco during twenty-four hours stay in Islay became indisposed, and when they had got to the city of Arequipa (about thirty leagues inland) no fewer than fifty-four of them were sent to hospital, of whom three serjeants died only a few hours after admission. This unfortunate event carried immediate panic into the hearts of the soldiery, so that, as the colonel says, not a few of the men were seized with intermittent fever and dysentery, for the cure of which he wished them to go to hospital, but instead of doing so, they betook themselves, all scared, to the mountains—following the road of Cucvillos to Lampa. The colonel remarks, " It was impossible to avoid our leaving behind at every resting place on the journey some absolutely grave cases, of whom, as well as of those left in the hospital of Arequipa, few have reached Cuzco."*

Dr. Tejada was the first medical commissioner sent by order of the Prefect of Cuzco to examine and report upon the nature of the fever which had broken out among the licensed soldiers of the battalion Cuzco at Sicuani—twenty-five leagues from the old capital of the Incas—where it first became epidemic in July 1855.† In a letter from Dr. Tejada, of date 12th July 1856, addressed to Dr. Garviso, and published in the *Resena Historica*, of the Cuzco epidemic, we have the following statement :—" The epidemic which reigned in Cuzco and in different provinces of the department, was constantly characterised by such symptoms as leave no doubt of its being the true yellow fever under different forms ; as from its first appearance in this capital, I officially intimated to the prefectoral authority on the 6th day of October 1855. In the deep and hot valleys it always carried the characteristic seal of the yellow fever—the typhus-icterodes or black vomit of authors,—while in the colder and elevated regions it assumed the character of typhus in its various modifications."† Dr. Bernardo Pacheco, as quoted by Dr. Garviso, arrived at the same conclusion as Dr. Tejada, with respect to climatic modifications of type observed by him in the adjoining department of Puno during the reign of the epidemic. The department of Cuzco is the most

* It is curious to contrast the fate of the battalion Cuzco, landed at Islay, with that of the portion of the battalion "Arequipenos Libres" which escaped shipwreck in the steamer *Rimac*, off the point of Lomas, on the 1st March, 1855. Of these troops, two hundred got safely on shore near Acari, and marched thence to Arequipa, about three hundred miles of desert, animated by the spirited example of endurance given by the daughter of their general, Albuzuri, without a single case of yellow fever. One chaplain and two or three men, overcome with fatigne, are said to have been the only victims of this march. Col. Narcisco Arestiqni ascribes much of the sickness in the battalion Cuzco to the fatigues of the comparatively short desert march from Islay to Arequipa, and supposes tercinna and dysentery to have sprung from fear of the pestilence which carried off so suddenly their three sergeants. But a better reason for the appearance of ague among their ranks would be that many of them travelled to Arequipa from the sea-port by the malarial valley of Vitor ; and, as for dysentery, it is the usual lot of strangers who drink of the crude water of Arequipa—though safe for culinary purposes—as shewn satisfactorily by Dr. Juan Gualberto Valdivia, in a discourse read at Arequipa on the 30th April, 1845, and published in his *Miselanea Chimica*.

† At the cold and lofty station Huancavelica, Dr. Villar describes the epidemic diseases of 1855-6 as in a great degree *masked* under the ordinary symptoms of typhus. *Lima Med. Gazette.*

—/ *Sicuani is at the elevation of 3,532 metres above the sea*
Tschudi

densely peopled in Peru, and on that account it suffered more than any other in this epidemic.

The city of Cuzco is situated at the elevation of 3,468 metres, according to Don M. E. de Rivero, in a transandine country, and lat. 13° 30′ south. And yet at this elevation, where the summer temperature in the shade is stated to be 12° or 13° of Reaumur, and the winter temperature from 7 to 8° of the same, yellow fever as described by the medical commissioners and others, raged with unmitigated virulence during the wet season (which is the summer) 1855-56. I have examined the statistical table of cases cured by Dr. C. Garviso in Cuzco, which is attested by D. Damasco Lechaga, the commissary of police, and find that out of one hundred and thirty cases reported in abstract, seventy-seven were of white race; twelve Mestizoes, and forty-one pure Indians. This enumeration shows that the yellow fever of Cuzco was not exclusively of a particular race. Dr. Garviso in his statistics mentions different modifications of the epidemic fever, under such headings as the icteroid, cyanose, bilious, petechial, adynamic-petechial, ataxic, and hæmorrhagic, &c.; under the denomination petechial-adynamic he relates the case of J. Vergara de Manca, of white race, seventy-five years of age, and of a weakly constitution, he says, " This woman was attacked at three o'clock in the morning, and instantly her body was covered all over with numerous black spots, the skin being of a yellow colour, and the prostration extreme. She had vomitings and cold sweats, with almost a glacial coldness of the body generally." This patient, we are told, was cured on the fifth day, by the help of ammonia and cinnamon, quinine, camphor, and opium. Another case—Rosa Rolanda, of white race, and thirty-five years of age, presented as symptoms, " Vomitings and evacuations of a bilious dark green appearance; the skin of an intense yellow with all the other signs and characteristics of yellow fever of the highest grade." Again, Josefa N——, an Indian or Chola, by race, is represented as having, " blackish-green bilious and sanguineous vomiting and purging with the yellow colour of the skin." Among the hæmorrhagic cases are enumerated a great variety such as, " nasorrhagias, gastrorrhagias, enterorragias, and abundant pneumorrhagias."

I cannot here occupy more space, or weary the reader's attention by a full description of all the horrors of the Cuzco epidemic of the wet season of 1855-56. By the church doors and open streets death suddenly seized its victims. The public hospitals and conventual asylums for females, were crowded to excess from November to May, with cases in every stage of progress. The newly attacked, the more aggravated, the dying, and the dead, were all seen huddled together. In January 1856, a very intelligent officer, Don Miguel Criado, who was one of the first attacked with yellow fever and black vomit in Lima (on the 14th May 1853), accompanied to Cuzco in capacity of Secretary, the new Prefect General, Buen Dia. The prefect was almost immediately on his arrival laid up with the fever, so that during his illness the daily duty of inspecting the different hospitals of the city fell on his secretary, who described that the pestiferous stench on entering these receptacles of human woe was overpowering—the air of the wards was saturated with animal miasms! He observed to me that the disease was the same in Cuzco as in Lima. Of two hundred convicts employed in Cuzco, in burying the dead, only two men escaped with their lives. (See *Note F.*)

Addendum. —At the port of Arica the licensed battalion Tacna, on its return to the capital and province of the same name, were disembarked on the 8th of April 1855, direct from Lima, and apparently well. They entered the town of Tacna on the 12th of the same month. They had barely taken up quarters when yellow fever broke out among them, in the suburban section called the Ranchería. They were supposed to have caught infection in Arica, where there had been previous to their arrival some cases of yellow fever. The fresh ingress of troops, therefore, supplied material for the epidemic.

Treatment of the Sierra Epidemic Yellow Fever.—We have filled our space and have no room left for a detailed account of medical treatment.

The Indians of Conchucos following a natural instinct, had a great craving for the "chulco," or wild cress, which they devoured with eagerness. In the fruit-growing valleys lime juice and iced lemonades were in liberal use.

In the valley of Chinchao, where it was feared that the fever would prove most fatal, matters turned out quite contrary to expectation. The malady was at once disarmed of its terrors, and arrested completely in its course by covering up the individual attacked with the fresh green and fragrant leaves of the indigenous erythroxylon coca. This novel appropriation of the staple production of the valley excited the most profuse perspiration, which ended in immediate recovery—the remedy was infallible. We can imagine how powerfully these leaves, full of empyreumatic qualities,

would affect the nervous susceptibility of the sick, and help
to invigorate and sustain vital action throughout the system.
It was observed with almost equal surprise that the
rum and brandy customers at the mines of Cerro-Pasco
escaped better than others in this epidemic. These alcoholic
drinks would readily supply materials for the production of
carbonic acid to be thrown off by the lungs. They would
thus support animal heat and prevent exhaustion. Dr.
Gallagher of the British Hospital, near Callao, found that
brandy and water freely taken all through the successive
stages of the fever, supported vital energy without ever
producing over-stimulation in the severe yellow fever of
1856.

Above all things, pure air, and separation of the sick from
the healthy, were found the great safeguards in preventing
the spread of the fever, and the *sine qua non* of successful
treatment.

It was a very usual practice to commence the cure of the
disease with laxative enemata and aromatic sudorific drinks.
Dr. Daza, early administered an emetic of ipecacuanha with
tartar emetic, which was followed up by an evacuant of
calomel and rhubarb combined, castor oil, or neutral salts.
This treatment tended to preserve the equilibrium among
the functions of circulation and secretion. Bleeding was
another resource with this physician at the outset of his
mission to Conchucos, when he bled largely and repeatedly
in Piscobamba—afterwards he bled with more caution in
Siguas; but latterly he abandoned venesection altogether in
the province of Lampa. To this southern district he was
recommissioned on his return to Lima from Conchucos.
On Dr. Daza's route to Lampa, where he arrived on the first
day of November, 1855, he saw thirty epidemized natives in
different gradations of the fever at the Indian village Chi-
guata, which is five leagues from Arequipa. He observed,
that all these sick had the same general symptoms which he
had been familiar with in Conchucos. We thus learn on the
most competent authority that the epidemic in north and
south Peru was of the same common nature.

Dr. Macedo believed, that in Huaras small bleedings had
been usefully employed by him in persons of a sanguine

tients acidulated drinks and ice, nitre and digitalis, whey and cooling enemata. Iced lotions, and the plug saturated with alumina or tannin, were applied to restrain excessive nasal hæmorrhage. When the stage of prostration had set in, he used pills, consisting each of one grain of camphor and one-fourth of a grain of opium, made up in extract of bark, for a dose: and to be repeated at intervals of two, three, or four hours, according to the degree of depression. Wine, soups, and a weak infusion of valerian were also administered. We are told, that by this tonic treatment many were cured. But in certain cases it was found requisite to deviate from this method, and he then used with advantage one drachm of Venice turpentine in four ounces of distilled water of orange flowers, made into an emulsion with gum arabic, of which a spoonful was given every two hours. He, likewise, had recourse to turpentine glysters made up with yolk of egg, which were found to remove abdominal pains and reduce the frequency of the pulse.

Again, in that non-hæmorrhagic, or congestive modification of the disease, called the "hepatic," by Dr. Macedo, when there was local pain or tumefaction observed in the region of the liver, this practitioner applied poultices, blisters, embrocations, sinapisms, turpentine epithems, or cupping-glasses as might seem most required. Besides, these topical applications, it was found advisable to adhibit a pill, consisting of three grains of calomel, and two grains of extract of rhubarb, repeated every two hours for twenty-four or forty-eight hours, with diluents, by which means the pulse became lowered and softer, and the general effect tranquillising. But in the stage of prostration of this hepatic modification, accompanied with delirium and coldness of the body, all remedies, we are told, were found to be almost equally useless, as death was generally inevitable. [See *El Restaurador de Huaras* of 11th November, 1854.]

After his return from Huaras, Dr. Mariano Macedo was recommissioned to the southern maritime department of Moquegua, of which Tacna is the capital. Whilst attending the sick of the epidemic yellow fever in this latter place, he experienced the great efficacy of the turpentine mode of cure in the hæmorrhagic form, having saved by it five out of eleven patients labouring under black vomit. But when this fatal symptom occurred in the congestive form, and during the stage of febrile reaction, he found turpentine of no efficacy. I may observe, that during strong febrile reaction in the hæmorrhagic form I have seen in my own practice, just as little success from this remedy.* It has been in the stage of greater or less prostration, attended with black vomit, that we have seen what may be fairly called resuscitations achieved by turpentine, especially in patients of pure or predominating Indian blood. Dr. Macedo, further employed in the hæmorrhagic cases, just as the second period was initiated, the capsicum, camphor and opium pills, as prescribed by Dr. Copland, conjoined with the infusion of valerian or arnica.† This reanimating treatment we have often seen of great service in supporting the vital forces, and preventing the gastric hæmorrhage, in which turpentine has been found so efficacious. In the lazaretto of Ayacucho, Espinosa never bled his patients, except in a few instances in which the brain or lungs appeared to be in a state of excessive congestion. He was partial to the use of rhubarb with calomel administered early in the malady in three successive doses, with a view to anticipate intercurrent hepatic congestion, which, according to his own observation during ten months attendance at the lazaretto, Quicapata, was not unfrequently developed in course of the fever, but never manifest in its earlier period. In the hæmorrhagic variety of the disease he derived much benefit from the cold bath, which brought on reaction and sanitary sweats. He also employed turpentine enemata and epithems to the abdomen with good effect.

The city of Arequipa is situated in an oasis in the midst of a dry sandy desert, at the elevation of about 7,800 feet above the Pacific. It was observed in this dry locality, as in the equally dry city of Piura in North Peru, that yellow fever conveyed from the seaboard did not propagate itself as a general rule   Many died outside in the suburbs, or on the arid plain and resting places (tambos) between the city of Arequipa and its seaport Islay. The periodical of that capital, " La Soberania," No. 35, has an article on the prodigies performed there by Dr. Adam Cridland with his " new treatment" of the epidemic. This consisted (as I have learnt from himself) of turpentine in large doses, suited to the necessities of the case. He observed, that in the invasion of the fever at Arequipa, head symptoms took precedence of the

---

* By referring to my notes, I find that this fact was remarkably illustrated in the case of the young Spaniard Francisco Ignacio Hurtado, who fell a victim to the disease in the Lima epidemic of 1854. I attended him in junta with Dr. C. Segura and Dr. Jose S. Corpancho.

† Lima Medical Gazette, vol. iii, No. 70. In one case of incipient black vomit, with great sensibility at the epigastrium, Dr. Macedo, in consultation with Dr. Wanne in Tacna, pushed mercurial treatment to salivation, and saved the patient.

gastric. In one remarkable case of a woman attended by him, the hæmorrhagic symptoms were so generally pronounced, that the blood filtered by the rectum, vulva, urinary passages, mouth, gums, nostrils, and punctæ lacrymaliæ. He applied ice to the spine, and administered as much as thirty-two drachms of turpentine mixed with gum syrup, before reaction commenced. When this change was established, he administered camphor and cinnamon with water and iced lemonades, by which means she recovered perfectly.

In Cuzco, bleeding appears at first to have been carried to excess, as it was reported that the epidemic of Sicuani was a catarrhal affection that was best cured by venesection and sudorifics. One worthy lady of my acquaintance, charitably paid two "sangradores," or bleeders on the occasion; but quaintly remarked, that both being dead of the epidemic, she did not find the poor died any faster than they did before. When Dr. Garviso arrived in Cuzco, he attacked the *modus medendi* of the Sangrado school. In his own practice, which he says was infallible, he discarded bleeding, blistering, emetics, and purgatives. In the milder cases, he used mild remedies, as laxative enemata and sudorifics. In the more severe cases, he relied on cold immersion and ergot of rye to restrain hæmorrhage. When needed, he supported the vital powers by tonics, antispasmodics and stimulants, such as phosphoric and sulphuric æther, camphor, cinnamon, tincture of valerian or castor, opium, ammonia, and quinine.

Pathological Remarks on the Sierra Yellow Fever.—One of the most general symptoms of this disease, was assuredly that yellow colour of the body, which gives this fever its distinctive name. To call it "bilious," is to assume that the icteric hue was the effect of a subordinate hepatic complication—a mere speciality. But that the yellowness was not dependent on hepatic lesion appears to me quite evident, when we reflect that it occurred in the more severe congestive as well as hæmorrhagic forms in Lima in 1854, wherein there was no hepatic inflammation or enlargement; that it was common to six out of the ten autopsies in Huaras with the five inspections in Tacna; and in seven out of ten in Huaras, the biliary ducts were free. Dr. Macedo* imagines that this outward appearance was peculiar to his "hepatic form" in the Sierra epidemic. We have shown from the observations of other practitioners, that it equally existed in

---

* *Lima Medical Gazette*, No. 51, p. 52, tom. iii.

the hæmorrhagic variety. In Tacna as well as Huaras, the
outward aspect of the body in the examples of *post mortem*
inspections, was not merely icteric, but dotted with mulberry
coloured spots. These concomitants of the yellow skin are
not intelligible on the theory of symptomatic jaundice. To
explain this subcutaneous extravasation, we must find some one
general principle or cause of disease independent of all minor
accidents and special complications. In those violent cases
where we have seen the victims cut off by a few hours illness,
and the cuticular exudation tinge the bed and bed-clothes
of the deceased, this effect could not be called hepatic, which
was caused with the rapidity of a thunderbolt. In the year
1828, Dr. William McLean of Lima was on the river Plate,
where he witnessed the cases of fourteen men struck by
lightning, of whom nine were killed on the spot, and five
recovered. The nine dead bodies had livid, or ecchymosed
spots on the skin; and the five living men had the skin as
yellow as it is ever seen in the most intense yellow fever. In
this scathing example, I find the only fitting ground of comparison with the fulminatory instances of death during the
period of invasion of the Sierra yellow fever, in which the
results were evidently more toxicological than anatomical :
the vitality and cohesion of the blood were at once assailed,
and the healthy concatenation of every function of the
system so suddenly subverted, that death necessarily
ensued.

When we view the effects of this blood-poison in its more
ordinary and less concentrated operation, and search for the
inward traces of organic alterations which the fever may
have left behind, the inquiry is not free of difficulties. The
living history, or successive events of the case before us, may
be imperfect or unknown, and the duration and character of
the primary fever will essentially modify its ultimate complications.

In a memorandum with which Dr. Daza, when a member
of Congress in 1860, favoured me, he says, " In Siguas, I
observed that the prevailing anatomical character of the
epidemic was a congestive state of the spleen ; whilst in the
colder temperatures of Puno, congestion of the liver was the
predominant symptom, with a remarkable alteration in the
duodenum. The mucous coat of this intestine presented a
dark vinous colour, and it was easily torn."

Dr. Pedro Espinosa of Ayacucho, was surgeon-in-chief of
President Echenique's army in 1854, attached to a force of
about five thousand men of all arms, encamped on the

heights of Chongos and Chupaca, in the province of Jauja, Dr. Espinosa had two hospitals under his inspection. As yet, the epidemic yellow fever was not seen in this district; but in all cases where perspiration was medicinally sought for, there was difficulty in procuring it on account of the coldness of the climate. The sick could only be made to perspire freely under a heavy weight of bed-clothes. Thus, the diseases prevalent among the troops—such as, acute articular rheumatism, pneumonia, dysentery and *tabardillo*, were mostly of a "congestive and inflammatory kind."* With these facts before us, we cannot be surprised to learn that the epidemic fever in the colder regions of the Sierra, was accompanied in its secondary effects, with local complications of a congestive or inflammatory nature. It is to be regretted that we have no *post mortem* illustrations of the pathology of this epidemic, either from Ayacucho or Cuzco. But Dr. Garviso specifies among the symptoms observed by him in Cuzco, a severe pain in the sacro-iliac and coccygeal region. This, we may suppose, would imply some grave affection in the great intestine, similar in some degree, to that observed by Dr. Destruge in the epidemic yellow fever of 1842-5 in Guayaquil. In a letter, kindly addressed to me on the subject of that equatorial pestilence, Dr. Destruge related, that when the Indians from the highlands of Quito descended to Guayaquil during the reign of infection there, and were taken ill and died of the fever, he invariably found on examination of their bodies after death, that the rectum was in a gangrenous state. He also observed that this affection of the large intestine was peculiar on that occasion to the mountain Indian—the mixed and dark races were free from it.

The alterations in the liver noticed by M. Louis in the Gibraltar fever of 1828, and which were very observable in Lima in 1854, many physicians look upon as the constant and characteristic anatomical conditions of this disease wherever it occurs. But this conclusion is at variance with facts, observed in Spain and Peru.

In the Lima epidemic of 1854, the post mortem state of the liver was always found dry or juiceless.† But in the Tacna yellow fever of 1855, we learn from Dr. Macedo's inspections, that the liver was only found "*dry*" in one case, out of five examined, while in another it was diminished in

---

* Dr. Espinosa's letter of 16th April, 1859, in the *Lima Medical Gazette*.
† *Lima Medical Gazette*, tom. iii, No. 51, October 1858.

D

firmness of texture, but increased in size ; and in a third
subject, it was found to be softened and congested, &c.
Here, then, at the comparatively low elevation of Tacna,
1816 feet, we see a great step made towards those hepatic
complications and climatic alterations, which the same praise-
worthy and laborious practitioner observed at the high elevation of Huaras ; and which also occurred at the still greater
height of Lampa and Puno, as noted by Dr. Daza.*   It has
been shown by the autopsies made in the lazaretto of Lima
(at the elevation of five hundred feet) in the year 1854,
that the intestinal mucous membrane presented spots of a
more or less crimson colour, but in the majority of cases,
as we are told by Dr. Macedo, without ulceration or softening of the membrane.   In Tacna, again, in 1855, the intestinal mucous membrane presented in some cases very signifiy
cant ulcerated points, and in every instance of five autopsies,
softening of the mucous coat of the stomach, and other equally
remarkable traces of degeneration of tissues in the large as
well as smaller intestines : similar effects were seen in
Huaras.

In Cadiz in 1800, in Medina Sidonia in 1801, in Malaga
in 1803, there were great variations as to the external symptoms of yellow fever—such as petechiæ, icteric colour of the
skin, and especially black-vomit, which last, Arejula did
not consider an essential symptom, but merely an accessory
fatal accident of the malady [Arejula, p. 145].   In Cadiz
the liver was found to be much enlarged and softened, and
of a colour passing from yellow to black.   In Medina
Sidonia the liver was observed to be clearer in colour than
in its natural state, but of normal size and texture.   In
Malaga the liver was seen of large size, and with jasper-like
brown spots, and of a firmer consistence than natural.   The
spleen also, was, in one case, found to be larger than in the
normal state ; and, in another case, this organ was entirely
destroyed by suppuration ; and the mucus coat of the stomach
ulcerated or gangrenous.   The lungs, in certain malignant
cases, presented gangrenous portions, which had caused fatal
hæmorrhages by the mouth.

Such facts as these, which I bring forward from the great
Spanish authority on the subject of yellow fever, clearly
evince that in Andalusia, during the specified epidemics,
there was no specific diseased state of the liver, which could

---

* By Rivero's calculation of barometrical heights of places in Peru, Lampa
is 3,901, and Puno 3,922 metres above the level of the sea.

be emphatically called the pathognomonic anatomical lesion of yellow fever.

In conclusion, Dr. Villar relates three cases of post mortem inspection practised by him during the epidemic of 1855-56, at Huancavelica in Peru, at the elevation of 3,798 metres (according to Ulloa as quoted by Rivero) of which the result was, that in all the three he found the stomach and intestines in a normal state, without any alterations in the follicles of Peyer nor of Brunner. The liver was hypertrophied extensively in two cases, one of which had the gall bladder full of biliary calculi, and in this, as in the other two, its consistence was slightly softened, but not so much so as the spleen. In the lungs the alterations were variable, as they were also in the brain—in one case of which, four follicles of pus were observed on the superior aspect. See *Lima Medical Gazette* of 30th Dec. 1857.

PATHOLOGICAL TABLE.

Drawn up from Dr. Don Mariano Macedo's published reports in the " Lima Medical Gazette."

It were to be wished that the columns confronted presented the same number of autopsies so as to render the comparison more complete; but they, each of them, respectively exhibit all the data in our possession on the necroscopical observations made in 1854, in Huaras, approximatively at the elevation of 10,000 feet, and in Tacna in 1855 at the elevation of 1,816 feet above the Pacific. The analysis of the ten Huaras *post mortem* cases is nearly word for word, as given by Dr. M. Macedo; the confronted analysis has been drawn up by myself from that physician's printed reports on the Tacna autopsies, only five in all.

OUTWARD ASPECT OF TEN CASES IN HUARAS.		OUTWARD ASPECT OF FIVE CASES IN TACNA.	
Yellow colour of the skin in	6	Yellow colour of the skin in	5
Out of the above six there were mulberry-coloured spots on	2	Livid or mulberry-coloured spots on the skin in the	5
Petechiæ disseminated in three cases out of the ten	3		
The Liver.		*The Liver.*	
Increased in bulk in	7	Less juicy than natural in	2
A little larger than natural	2	Dry	1
Of natural size in	1	Somewhat softened and gorged with dark liquid blood	1
The parenchyma reduced to pulp	7		
The parenchyma softened	3	Somewhat increased in size, softened, and easily torn in	1
The colour of the liver was more or less a dark mulberry, and		The colour of the organ was of a	

spotted with capricious jasper-like figures, in - - - - - - 7
Of a darker colour than natural in 3

Gall-Bladder.
Dilated in - - - - - - - - 6
Of normal size in - - - - - - 4
Biliary ducts free in - - - - - 7
But narrowed in - - - - - - 3

Colour of the Bile.
Green in - - - - - - - - 7
Of the appearance of black pitch - 1
Colour of coffee - - - - - - 1
Colour of *chicha*, (or turbid beer) 1

The Spleen.
Increased in bulk in - - - - - 7
Of natural size and consistence in 3
Reduced to a state of pulp in - - 7

Stomach.
The stomach dilated to more than one-third its normal dimensions 6
Stomach of natural size in - - - 4
The mucous membrane entirely softened in - - - - - - - 8
Mucous coat less consistent than natural in - - - - - - - 2
The mucous coat presented on a ground more or less of a leady hue, mulberry-coloured or crimson spots, and was sprinkled over with red points or specks in 10
Small ulcers of various sizes - - 4

Contents of the Stomach.
This organ contained a dark green liquid in - - - - - - - - 3
A liquid like turbid beer, or *chicha* 1
A liquid like the lees of wine in - 1
A liquid like coffee with milk in - 1
A free liquid of a yellow hue in - 4

Small Intestines.
Viewed externally, these presented more or less slate coloured bands, with vascular arborescent ramifications, in - - - - - - - 10
At different points of the mucous membrane there were mulberry

light rhubarb, or dried rose shade, with red, livid, or mulberry spots, and irregular arborescent jasper-like figures in - 5

Gall-Bladder.
A little dilated in - - - - - - 1
Less size than natural - - - - 1
Biliary ducts free in the - - - 5

Colour of the Bile.
A dark green in - - - - - - 5

The Spleen.
Congested with black blood in - 1
Somewhat congested - - - - 1
Natural state in - - - - - - 1
Of natural size, but softened, and breaks easily on being handled - - - - - - - - 1
Of small size and reduced to pulp 1

Stomach.
The stomach dilated from one-fourth to one-third its natural size in - - - - - - - - 2
Dilated to twice its natural size in 1
Dilated, but to a less degree, in - 2
The outward aspect of this organ was usually lead-coloured, and showed in some cases vascular ramifications.
Viewed internally, the mucous membrane was completely softened in - - - - - - - - 4
Its consistence diminished in - - 1
The mucous coat partly leady-coloured, and partly of a dark red, with numerous red specks - - 2
The mucous coat with accuminated red specks disseminated - - - 3
Dark coloured patches on the mucous membrane, and varnished with a muco-purulent matter - 1

Contents of the Stomach.
This organ contained a dark coloured sanguineous liquid in - 3
A thick liquid the colour of coffee 1
A liquid the colour of soot in solution - - - - - - - - 1

Small Intestines.
General appearance of the mucous membrane, from a dark leady to an ash colour, in - - - - - 4
The mucous membrane presents red or dark reddish spots in different parts in - - - - - 3

and leady-coloured spots in - - 10

In different places along the intestinal mucous membrane—and very particularly in the ileum—were seen minute ulcers grouped in bands on an ash-coloured ground in - - - - - - 5

The mucous coat entirely softened at different points in - - - - 9

Id. less consistent than natural - 1

Red spots meeting in arborescent outline - - - - - - - 1

Over the whole extent of the small intestines there is an eruption of small papillæ of an ash colour, but especially occupying about one yard in extent of the lower portion of the ileum; and, besides there are longitudinal bands on the same portion of the ileum, set with minute ulcers on an ash-coloured ground* - 1

The mucous membrane ulcerated and softened, so as easily to give way under the pressure of the nail in - - - - - - - 2

The mucous membrane softened, or less consistent than natural - 1

Large Intestines.

Red spots and ulcerated points of different dimensions—the larger of six, and the smaller of two lines - - - - - - - - 2

Numberless ulcers, the smaller of four lines and the larger an inch —which latter *in one case* resemble dysenteric ulcers, with softened mucous coat, and lubricated by a thick dark fluid - 6

A thick sanguineous liquid in the ascending portion of the colon; the mucous coat of a reddish mulberry colour, with blackish transverse bands which looked like gangrenous scars, but without the characteristic odour; and ulcers at certain points with softening of the mucous membrane 1

Large Intestines.

The mucous membrane of a dark leady hue, and at intervals presenting red arborescent stains, is notably softened and easily torn, ulcerated, and varnished with a thick leady-coloured fluid 1

The cæcum thickly coated with a muco-sanguineous matter like quince-jelly; and the remaining large intestines are covered with a thick sanguineous mucus, and the mucous membrane less firm than natural - - - - - - 1

The mucous coat with a sanguineous exudation in - - - - 1

The larger intestines without any fluid, but with the mucous membrane less consistent than ordinary in - - - - - - - - 1

The mucous membrane varnished with a dark liquid, but no other lesion, in - - - - - - - 1

Contents of Intestines.

The intestines varnished in different places with a dark green liquid of different shades of obscurity - - - - - - - - 5

Varnished with a yellowish mucus 4

With a dirty fluid in - - - - - 1

Contents of Intestines.

The mucous coat of duodenum and other small intestines coated with a thick dark green liquid in 1

The mucous coat varnished with a dark fluid - - - - - - 1

Do. varnished of a leady hue - - 1

Do. varnished of the colour of soot in solution in - - - - - - 1

The duodenum and part of the jejunum varnished with a thick liquid the colour of coffee - - 1

The above are, in abstract, the most remarkable anatomical

---

* *N.B.* These peculiar ulcerated bands on the ileum, with the special eruption noticed above, occurred in the case of Serjeant Villalba, who died on the third day of congestive yellow fever, with black stools and black vomit.

alterations observable in Dr. Macedo's *post mortem* inspections.

Both at Huaras and Tacna the few cranial inspections recorded, show that in several cases the dura mater was tinged yellow, that the sinuses and vessels distributed on the meninges were more or less loaded with blood, that the cerebral mass was more or less altered in consistence, and more or less dotted with red specks. In Tacna the valves of the heart were yellow, and its cavities contained dark fluid blood in four cases; in Huaras the valves of the heart were very yellow in one case, and in its cavities contained a yellow coagulum in two cases. The lungs in Tacna were livid in one case, and in Huaras partial hepatization and congestion occurred in one case. The texture of the kidneys sound in Tacna and Huaras. In Huaras the pancreas was sound in the ten cases, and in Tacna the pancreas is passed over in silence, which implies its soundness. In Huaras the urinary bladder was empty and contracted in five cases, dilated in one, and with the mucous coat sound in the ten. In Tacna the urinary bladder was contracted in two cases, and more or less distended in three; mucous coat of a yellowish colour but healthy texture in four, and of a yellow tinge, and dotted with red points in one. The urine was more or less discoloured or yellowish and turbid, in both Tacna and Huaras, in five cases.

I offer no comments on the numerous points of agreement in the above enumeration of morbid appearances among the cases examined in Tacna and Huaras. They are obvious at first sight; and wonderfully corroborative of the elective action of the morbid agent on the texture of the stomach and intestines, notwithstanding the great elevation of full eight thousand feet, and nine degrees of latitude, that separate the one place from the other.

––––––––

NOTE A.

An epidemic, which much resembled the " mattazahuatal," broke out in Huamanga and Cuzco in 1719.20. In 1759 another epidemic, more like the milder forms of yellow fever, but of which our present limits will not allow us to treat at large, spread from the mountains to the coast with great rapidity, affecting domestic animals as well as the human race. (See Ulloa's *Noticias Americanas,* p. 202 3.)

NOTE B.

Lower down than Caras, in the warm and sugar-growing district of Huaylas, is the estate of Pumacocha, the property of Don Ignacio Figuroa, in 1860 a deputy for Huaras and senator in Congress. This gentleman gave me the following particulars of the epidemic of 1854, as observed by him, on this estate of Pumacocha:—

"In the majority of cases, the invasion took place in the morning with headache, perturbation of mind, and pain at the stomach. By evening, the skin assumed the yellow hue, and then the great agitation ceased; patients thus affected remained quiet and complained of nothing, but on the morning of the following day they were dead, some of them voiding blood by mouth and nostrils, and others not. There was no black vomit, but some vomited matter of the colour of *dark coffee*. Out of one hundred and fifty individuals seized with this fever on the estate of Pumacocha, only thirty escaped with life. The negros, both male and female, were the last to be attacked, and none of them died of the epidemic."

NOTE C.

After the most careful consideration of this evidence, the following is the conclusion arrived at by the late lamented Dr. J. O. McWilliam, as communicated to me in a note from himself, dated November 9th, 1861:—

"It is the most stupendous instance we possess of the yellow fever having (as now begins to be acknowledged) extended beyond the limits assigned to it by Humboldt and almost all subsequent observers."

NOTE D.

Dr. P. Espinosa considers the epidemic, as seen by him in Ayacucho—at an elevation, probably, of about 9,000 feet—as a hybrid between the yellow fever and Tabardillo. As our *typhous* fevers are divided into typhus and typhoid, so the oldest practitioners of Peru, thirty years ago, distinguished their Tabardillo into simple *Tabardillo*, and the *Tabardillo-entripado*, which latter meant their *enteric* form of typhous fever.

For information of the usual modifications of type in the fevers of Peru, other than is given at p. 311, see my paper in the *Edinburgh and Medical and Surgical Journal*, vol. liii.

NOTE E.

It is worthy of record, in illustration of the facts mentioned, p. 312, that a year before this, Dr. Rosas, a vocal of the court of Arequipa, was, on his return from Lima, laid up with yellow fever at Islay. His colleague, Dr. Segar from Tacna, with his whole family and domestic establishment, to the number of sixteen or eighteen souls, were next seized with the same fever, some in a mild and others in a more severe manner; but Dr. Segara himself was taken ill on the 21st of April, and died on the 8th of May, 1854, with black vomit. It was thought this gentleman had caught infection from visiting Dr. Rosas during his illness.

NOTE F.

I was told by Don Miguel Criado that the dead bodies he saw in hospital at Cuzco were of diverse colours, such as yellow, blue and yellow, and also violaceous or livid, mottled, etc. This gentleman observed that in all cases the epidemic commenced with intense headache; and that in some, vomiting came on in the invasion, when the matter ejected soon passed from yellow to a mahogany colour, and finally to a dark green or black. Col. Aranzabal, who seeing the fatal effects of bleeding in Cuzco devoted himself to the hydropathic method of cure, confirms Senor Criado's observations, and adds, that the feculent motions often looked like "brea," or tar, and that the urine was of the colour of dark coffee.